GREAT ZOOS OF THE WORLD

GREAT ZOOS OF THE WORLD

THEIR ORIGINS AND SIGNIFICANCE

EDITED BY
LORD ZUCKERMAN
OM, KCB, MD, DSc, FRS

President, The Zoological Society of London

Routledge
Taylor & Francis Group
LONDON AND NEW YORK

First published 1980 by Westview Press

Published 2018 by Routledge
52 Vanderbilt Avenue, New York, NY 10017
2 Park Square, Milton Park, Abingdon, Oxon OX14 4RN

Routledge is an imprint of the Taylor & Francis Group, an informa business

Library of Congress Catalog Card Number 80 50412

ISBN 13: 978-0-367-02241-9 (hbk)

Contents

Acknowledgments

Between pages 56 and 57:

Blijdorp Zoo, *Diergaarde Blijdorp (phot. Peter Stephen Molkenboer).*
La Grande Rotonde, *Service Photographique du Parc Zoologique, Paris.*
Elephant House, *Zoologischer Garten Berlin.*
Bombed Elephant House, *Zoologischer Garten Berlin.*
Versailles, *Giraudon.*
Goma, *Photo Lang, Zoologischer Garten Basle.*
Paddiwack, *Camera Press (phot. Colin Davey).*
Indian rhinoceroses, *Zoologischer Garten Berlin.*
Engraving of London Zoo, *Michael Holford.*
Lion Terraces, *Zoological Society of London.*

Between pages 88 and 89:

Przewalski's horses, *Archiv Zoo Praha.*
Pheasant, *Jersey Wildlife Preservation Trust (phot. Phillip F. Coffey).*
Manchurian cranes, *Ueno Zoo.*
Wisent, *Photo V. Six – Zoo Anvers.*
Okapi, *Photo V. Six – Zoo Anvers.*
Somali asses, *Zoologischer Garten Basle (phot. Elsbeth Siergrist).*
Margay, *Arizona–Sonora Desert Museum.*
Surmatran tiger, *Frank W. Lane.*
Cheetahs, *National Zoological Gardens of South Africa.*
Fruit bat, *Jersey Wildlife Preservation Trust (phot. Phillip F. Coffey).*
Maned wolf, *Tierpark Berlin (phot. Gerhard Budich).*

Between pages 120 and 121:

Pelicans, *Tierpark Berlin (phot. Gerhard Budich).*
Slimbridge, *J. B. Blossom, Wildfowl Trust, Slimbridge.*
New York, old aviary, *New York Zoological Society.*
Bird House at Frankfurt Zoo, *Zoologischer Garten der Stadt Frankfurt-am-Main (phot. Ernst Müeller).*
Snowdon aviary, *Zoological Society of London.*
Bears, *St Louis Zoological Park.*
Rock at Peking Zoo, *Robert Harding.*
Antelope Yards, *St Louis Zoological Park.*
Wolf woods, *Philadelphia Zoological Gardens (phot. Franklin Williamson).*

Totem poles at Washington Zoo, *Max Hirschfeld.*
East Berlin Zoo, *Tierpark Berlin* (*phot. Gerhard Budich*).
Aviary and waterfall, *Jurong Bird Park.*

Between pages 152 and 153:

Nyala antelopes, *Zoo Praha* (*phot. Jiri Volf*).
Eaglet at Washington Zoo, *Max Hirschfeld.*
Koalas at Taronga Zoo, *Frank W. Lane* (*phot. Garth Grant-Thompson*).
Giant pandas at Peking Zoo, *Robert Harding.*
Asiatic enclosure, *New York Zoological Society.*
Nairobi Village, *San Diego Zoo* (*phot. F. D. Schmidt*).
Indo–Malayan Pavilion, *Metro Toronto Zoo.*
Polar bears, *Metro Toronto Zoo.*
Monorail and Indian rhinoceroses, *New York Zoological Society* (*phot. Bill Meng*).

Foreword

When I started my professional career as a junior research worker of the Zoological Society of London, I was told that zoos made news 'second only to murder'. Even if they do not still enjoy that exalted position in the hierarchy of public interest, they certainly continue to receive their fair share of attention. Births of rare creatures attract almost as much notice as royal births, and the death of a zoo favourite is treated as a national – sometimes even as an international – event. The escape of a dangerous animal from a safari park, and the mauling of a visitor or keeper, make front-page news. And when a book that attacks zoos appears, it receives almost as much attention as a major work of literature or science.

The present volume is neither a defence of zoos nor a guide to them. What it does is illustrate how zoos have evolved, by reference to the history of a few. Those that have been selected are among the best in the world. Needless to say, there are others that might have been chosen.

The names of the present and former professional staff of the Zoological Society of London who are responsible for the different chapters are listed on page 2. Most of us have seen the majority of the zoos that are described; and all have seen more than a few. I make no apology for the fact that the Zoological Society of London is treated as a standard by which to judge other institutions of a like kind. That is what it happens to be. It is the oldest and the most scientific of them all, and I like to think that it will continue to be the most vigorous.

The avoidance of repetition and the coordination of the stories which the various chapters tell has called for a liberal measure of general editing, which in turn has demanded an unusual degree of tolerance from my fellow contributors. I owe a special word of thanks to Mr Peter Olney, the Society's Curator of Birds and the present Editor of the *International Zoo Yearbook*, for his help in organizing the work.

S. Zuckerman
September 1979

The Contributors

Assistant Editor

P. J. S. Olney, BSc, Dip Ed, FLS
Curator of Birds, Editor of the *International Zoo Yearbook*,
Zoological Society of London

M. R. Brambell, VetMB, PhD, MRCVS, FLS
Director of Chester Zoo

Marcia A. Edwards, PhD, FLS
Editor of the *Zoological Record*, Zoological Society of London

C. G. C. Rawlins, OBE, DSc,
Director of Zoos, Zoological Society of London

J. W. Toovey, AA Dipl (Hons), FRIBA
Architect, Zoological Society of London

The Rise of Zoos and Zoological Societies

Lord Zuckerman

No one knows exactly when or where zoos first appeared on the scene. What we would be inclined to call zoos certainly existed in Ancient Egypt, but by then they had already become so full-blown, so much part of a codified form of life, that it is hardly possible that they did not develop from something more primitive. The word itself is of relatively recent origin. According to the Oxford Dictionary, it dates only from 1847, when it came into use as a colloquial abbreviation for the Zoological Gardens in London. It was obviously an attractive word, since it soon spread to other countries and into other languages – so usurping the earlier term 'menagerie', whose derivation provides a useful lead to the history of the keeping of exotic animals in captivity.

The word 'menagerie' comes from the French *ménager*, a term which comprehends a number of activities related to the management of a family and the care of a household. Until the middle of the seventeenth century, a menagerie was merely an enclosure, presumably part of a château, where livestock was kept and fattened for the table.[16] Louis XIV then decided to transform the menagerie which was attached to his palace at Versailles into what today would certainly be called a zoo. Almost immediately his example was followed by other landowners who either kept or who wanted to keep wild animals, partly out of curiosity, and partly as a mark of their station. Before that time there was no general term to denote an enclosure for wild beasts.

THE FIRST PHASE – REVERENCE AND BRUTALITY

The keeping of animals, however – either because they were objects of worship or as a mark of social status – goes back much further than Louis XIV.[13,16] The ancient Egyptians held certain animals sacred (for example the lion, the baboon and the ibis); they protected captive specimens in temples and parks; and they mummified the remains of animals and kept them in specially built mausoleums. They dreamt up mythological beasts; pharaohs and ptolemies sent expeditions abroad to bring back to Egypt exotic creatures; they set aside parks in an effort to 'acclimatize' species other than the cow, sheep, horse, dog and cat – which had been domesticated as far back as the Neolithic Age.

Hundreds of wild animals were used in Egyptian triumphal processions; camels, elephants and creatures such as the lion and the oryx were trained to pull chariots. The pharoahs of the eighteenth, nineteenth and twentieth dynasties engaged in royal hunts which, it is said,[18] did not have 'the remotest connection with ideas of food production, food magic, or protecting the flocks'. They trained cats, cheetahs, even leopards and lions for the chase, and hunted just for the fun of it. Many if not all aspects of the Egyptian cult of wild animals appear in the records of other ancient civilizations. There were certainly collections of captive animals in Greece by the seventh century BC, and by the fourth century BC they probably existed in most, if not all of the Greek city-states. Aristotle, who wrote a celebrated *History of Animals*, must surely have visited zoos; his most famous pupil, Alexander the Great, sent back to Greece many creatures which were captured on his military expeditions.[7] There were also 'zoos' in Mesopotamia, in India, and as far afield as China, where one, called the Garden of Intelligence, was set up, presumably for scientific and educational purposes, in about 1100 BC. As late as the end of the thirteenth century, Marco Polo records that he saw lions and tigers wandering freely through the rooms of a Chinese imperial palace.

During the lengthy opening phase of the evolution of zoos, the cult of the exotic animal also flourished in Ancient Rome, which at least as early as the third century BC had become acquainted, by way of travelling circuses, with a variety of strange animals. But it did so in a much more exuberant and violent form than in Egypt, where for thousands of years concern with wild animals was usually associated with reverence and mysticism. In fewer than one thousand years the Romans succeeded in transforming their interest in exotic creatures into a depraved and brutal cult.[16,19,24] They soon outdid the Egyptians in triumphal processions. There were schools in which professional gladiators perfected their skills and, as in Egypt, establishments for the taming and training of animals, in which baboons were taught to play the flute and zither, or pick out the different characters of the alphabet on command.[16] The Romans needed more and more lions, tigers, elephants, bears and strange deer to parade before the public. Proconsuls in distant parts of the Empire wanted them not only for their own displays; they also had to arrange for their capture and despatch to Rome. In all the many amphitheatres of the Empire, and in the Coliseum in Rome in particular, the slaughter of wild animals became an integral part of gladiatorial contests.

Twentieth-century man has become inured to the fact that millions of his fellow beings have been killed in war, that genocide has been practised on a hideous scale. But it still seems strange that for centuries people crowded into vast stadiums to watch not just two professional gladiators in mortal hand-to-hand combat, but, on occasion, pair after pair fighting to the death in a continuous show that sometimes went on for months. To celebrate his Dacian victory in the year 106, Trajan held games for four months on end, in the course of which, so record has it, 10,000 gladiators perished and 11,000 beasts were killed. Some fifty years before, Nero pitted a company of horsemen against 400 bears and 300 lions. The arena of the Coliseum was sometimes turned into a lake on which gladiators in boats tried to kill hippopotamuses, seals and crocodiles. Professional animal-killers battled bare-handed with bears and tigers and lions before frenzied crowds, with a far greater chance of survival than the untrained Christians – just one of several 'criminal' sects who were 'thrown to the lions'. This happened in amphitheatres not just in Italy, but in the furthest corners of the Roman Empire.

Most gladiators were prisoners of war or criminals. But there were also amateurs who were keen to display their skills in hand-to-hand combat – even, it is said, senators and 'women of high birth'. And they too fought not just between themselves, but with wild animals. Small wonder that the emperors, the senators and proconsuls of Ancient Rome needed their menageries. Between 1 BC and AD 14 Augustus kept – which presumably means that most were killed in the arena – 3,500 wild beasts, including a 'serpent' that was 82 feet (25 metres) in length. In the three years he ruled, between AD 79 and 81, Titus kept, that is to say 'used up', 5,000. To maintain the slaughter, there had to be menageries in which ferocious beasts were confined; there had to be parks in which the more peaceful grazed.

Gladiators symbolized man's mastery over beast. Those who survived, those who were more skilled in killing, were adulated as public heroes; their favours were sought by fashionable women; they were rewarded as no football players or prize-fighters are today. And they were immortalized in sculpture. The cult of the gladiator continued long enough to generate an everlasting sense of awe and admiration for men who behave as though they have no fear of wild animals.

It took centuries before gladiatorial practices came to an end. While there were a few prominent voices which protested against the contemporary taste for human and animal butchery, it was not until the year 325 that Constantine, the first Roman emperor to become a Christian, decreed that gladiatorial games had to stop. Unfortunately, it was not an order that could easily be implemented. Indeed, a contrary decree was issued in the sixth century by Justinian, the most famous of all the emperors of the Eastern Roman Empire and – so history tells us – a scholarly and serious-minded man. Justinian had taken first as mistress and then as wife, Theodora, the daughter of a keeper of bears in one of the two amphitheatres of Constantinople. Before she became his consort, Theodora used to perform, both in the arena and on the stage, in highly seductive parts, rather like a present-day striptease artist. She clearly saw nothing wrong with arenas, and no doubt it was at her behest that after their marriage Justinian passed a law which charged consuls to resume gladiatorial and animal combats, with the strict instruction that priests and bishops should not be allowed to be spectators. Animal contests were still being held in the amphitheatre of Constantinople in the twelfth century.

THE SECOND PHASE – THE MARK OF SPLENDOUR

Man's fascination with wild animals certainly did not come to an end when its expression became less sanguinary. Indeed, a second phase in the evolution of zoos may be regarded as having started when organized gladiatorial games stopped. Centuries before, the pharaohs and Ptolemies had craved exotic beasts as gifts, and King Solomon in his time treasured the apes and peacocks sent him as tribute every three years. Exotic animals went on being regarded as gifts at least as precious as the plate and jewels that princes and kings gave one another. In about AD 900 Harun al-Rashid, the most celebrated of Arabian caliphs, and also a hero of the *Thousand and One Nights*, sent Charlemagne an elephant which is said to have been the first to be seen in France (although some Gauls must surely have seen Hannibal's elephants on his famous march from Spain to the Alps in 219 BC[5, 22]). In the early part of the twelfth century Henry I established a small zoo at Woodstock, near Oxford, many of his animals having been given to him by other monarchs. In the thirteenth century Frederick II, King of Sicily and Jerusalem, kept an extensive menagerie. His animals – which accompanied him on his foreign travels – took part in a grand procession at Worms on the occasion of his marriage in 1235 to Isabella, daughter of King John of England and sister of Henry III, the monarch who moved to the Tower of London the zoo at Woodstock which his grandfather had established. There the English royal collection, such as it was, remained – with its upkeep for a time paid for by the sheriffs of the City of London – until the Zoological Gardens in Regent's Park were opened in 1828. Record has it that there were zoos in Mexico in pre-Columbian times. Dresden had a princely zoo in the sixteenth century. Philip VI of France kept lions and leopards at the Louvre in the fourteenth century in a building called the Hôtel des Lions du Roi. His successors on the throne of France seem to have kept menageries wherever they built palaces. When, after the outbreak of the Thirty Years War, the seat of the Holy Roman Empire moved from Prague to Vienna, the Royal Zoo moved with it.

It was not only zoos which survived when gladiatorial contests stopped. Circuses flourished, even if one may doubt whether their performing animals were as well trained as their counterparts are said to have been in Ancient Egypt or Rome. Vestiges of a more brutal, less humane fascination with animals still persist. Bullfights and cockfights continue. Bear-baiting has probably not died out in some parts of the world, and dancing bears and organ-grinders' monkeys can still be seen. Our television screens show us trained chimpanzees dressed in human clothes advertising some household commodity. The tricks of the modern circus – the acrobat and the trained animal, the lion-tamer with his whip, and the ringmaster – all these are relics of a more violent past, in the same way as the princely private menageries that were retained well into the eighteenth century reflected the grandeur that was the mark of the pharaoh. Some still admire the 'machismo' of the man who can frolic with tigers or who can put his head in a lion's mouth. And a sense of reverence for animals also still exists. Many people still solemnly bury their pet dogs and cats and erect headstones over their graves. When Guy, the celebrated thirty-year-old gorilla of the London Zoo, died in 1978, there were those who wrote to the Society objecting to his remains being used for scientific study, and to his skin being mounted for exhibition in the British Museum of Natural History. They felt that he should be given a proper burial. Heads of state are still presented with rare creatures. Only a few years ago, the London Zoo felt constrained to agree to a request of a lord mayor of London that it provide a llama cart to take part in the traditional triumphal procession on the occasion of his accession to his new office. He had, in fact, asked for camels. And those who knew beleagured Britain during the Second World War and the harsh years of reconstruction which followed, will surely remember pictures of Winston Churchill at the London Zoo looking with pleasure at his lion, Rota, no doubt rightly seeing in the beast a reflection of his own indomitable spirit.

THE THIRD PHASE – SCIENCE AND EDUCATION

Other than the Chinese zoo of 1100 BC to which reference has been made, and Ptolemy II's reputed zoo in Alexandria some 800 years later, there is little sign that zoos were regarded as institutions which had a part to play in the advancement of science and the spread of education before the Versailles menagerie was established in 1665. Until then their obvious purpose was the gratification of curiosity and the underlining of the magnificence and power of their owners. There are few indications that early students of comparative anatomy obtained the specimens which they dissected from the zoos of their times. Most of the creatures which they studied belonged to species that were readily available, specimens of which could be dissected 'fresh', and not, as they would be today, injected with preservative fluids – animals such as cats, tortoises, vipers, pigs and dogs. The researches which led William Harvey to an understanding of the circulation of the blood were made principally on deer. Kenneth Rose[21] tells us that: 'No creature was secure from Harvey's scrutiny, and his published works reveal a familiarity with more than sixty species.' In one of his 'characteristically limpid notes on comparative anatomy', Harvey writes:

> I have also observed that nearly all animals have truly a heart, not the larger creatures only and those that have red blood, but the smaller and pale-blooded ones also, such as slugs, snails, scallops, shrimps, crabs, crayfish and many others; nay, even in wasps, hornets and flies I have, with the aid of a magnifying glass and at the upper part of what is called the tail, both seen the heart pulsating and shown it to others.

It is true that more exotic creatures than these were used by some scholars, but in most cases their source remains a mystery. For example, one would like to know how Tyson, a London physician and scientist of the latter part of the seventeenth century, and his colleague Cowper, both renowned in the history of comparative anatomy, obtained some of the specimens which they dissected – among them the rattlesnake, the opossum and a chimpanzee that had died shortly after arriving in London after a lengthy journey from Angola.[17] And how did the elephant which Patrick Blair dissected in 1706 come to die near Dundee, and by whom had it been owned?[2]

Moreover, up to the end of the seventeenth century, most anatomical studies of exotic animals seem to have been undertaken in an unsystematic way, inspired more by curiosity about what would be discovered than as a test of some general theory about animal physiology or relationships. It was only when major biological generalizations started to be formulated that a new meaning was given to zoos.[18] This began in 1735, when the Swede Linnaeus launched his revolutionary ideas about systematizing the classification of plants and animals. Then there was the Frenchman Buffon who, while opposed to what he regarded as Linnaeus's simplifications, formulated the concept that all in time and space, a concept which he, in association with an anatomical living forms are related to each other in a single extended order, both colleague, Louis Daubenton, then illustrated in a vast work which occupied him from 1749 to 1767. In 1739 Buffon was appointed keeper of the Jardin du Roy which, as the Paris chapter relates, became in due course the Jardin des Plantes, and so the home of the national zoo of France. Through its association with the Muséum d'Histoire Naturelle, the zoo in effect became for a period the centre of biological research for France.

After Buffon there was Georges Cuvier, the man who helped Napoleon to introduce his educational reforms, and who, because of his skill and achievements as a comparative anatomist, is regarded by many as the founder of modern comparative anatomy. It was Cuvier who argued the principle that all structures in an animal are correlated. When carried to extremes this view implies that it should be possible to infer from the shape of a single bone the nature of the rest of the animal's anatomy, and so its habits and its place in the scheme of things. As a professor at the Muséum, Cuvier could hardly have failed to use the Zoo as a laboratory. There was Lamarck, a profound scholar and friend of Buffon, who was also concerned to understand the causes of the diversity of living and fossil creatures. Somewhat unfairly, his views are today usually remembered only in the simple catch-phrase 'the inheritance of acquired characters'. There was Buffon's German near-contemporary, Alexander von Humboldt, another late eighteenth- and early nineteenth-century scientist of wide interests, who promulgated his own views about a universal natural order. All these men, as well as other contemporary scholars, realized that there was far more to a zoo than a place where the uninformed could gaze in wonder. What is more, some were powerful men of affairs, and were able to make their voices heard.

This third phase in the evolution of zoos can therefore legitimately be called the scientific phase. It started in Paris and London. In France it came about formally as an unpredictable consequence of the overthrow of the monarchy. The animals in the King's menagerie in Versailles, together with those from travelling circuses, were moved to the Jardin des Plantes in Paris, and so scientists in the Muséum d'Histoire Naturelle were provided with a counterpart museum of living animals, and the people of France with a garden from which they could not be excluded. Both the Muséum and the Zoo were public institutions; they were on the same site; and the professors of the Muséum had immediate responsibility for the care of the animals when alive, and their further study when dead.[6,15]

The course of events in London was different. As related in that chapter, the first step was taken in 1826 with the founding of the Zoological Society, one of a number of societies that were being launched at the end of the eighteenth and in the early part of the nineteenth century to cater for specialized areas of scientific concern. When the Society was granted its own Royal Charter in 1829, its purpose was specifically stated as being 'the advancement of Zoology and Animal Physiology and the introduction of new and curious subjects of the Animal Kingdom'. From the outset the Society, the first of its kind in the world, was thus a scientific body, and the establishment of a zoo, to which the remains of the British royal collection were sent from their miserable dens in the Tower of London, was, at least in theory, one of the means whereby the Society could discharge its scientific and educational responsibilities. While at first a private 'garden', the Zoo was not an undertaking that was launched for the beguilement of those members of the Society who were not scientists. Today, any departure from the Society's purpose as set out in the Royal Charter would endanger its status as an educational charity.

Not long after the London Zoo came into being, many of the major cities on the Continent started to follow suit. There were still private princely collections, such as the zoo at Schönbrunn in Vienna, and at the end of the century great land-owners were starting collections. The most illustrious was that of the eleventh Duke of Bedford, whose motive was not just the traditional embellishment of a private park with deer and rare cattle, but the conservation of threatened species. It was he more than anyone else who prevented the Père David species of Chinese deer from becoming extinct. The last wild members of this species roamed the grounds of the Forbidden City in Peking, and were slaughtered at the turn of the century during the course of the Boxer Uprising against foreigners and Christians, a 'revolution' which also led to a diminution in the power of the royal house of China. The Duke of Bedford managed to obtain a few specimens of this otherwise extinct deer from some Continental zoos to form the nucleus of what became the famous Woburn herd. Together with other European land-owners he also played a significant part in the preservation of the European bison – the wisent.

In the early part of the nineteenth century a demand arose for zoological gardens that were open to the public by right, or at least to a company of subscribers or to people who paid for admission. Among the European cities which by 1860 had launched their own zoos were Dublin, Bristol, Berlin, Frankfurt, Antwerp and Rotterdam, and before the turn of the century their example had been followed by many others, including, as the movement spread, some in the United States and Australia. Most were founded by prominent citizens who felt that no great city should lack a good zoo. Some were started by men who collected and dealt in exotic creatures, and who also ran circuses. Enthusiastic amateurs played their part. Rotterdam Zoo was launched by two railway workers who kept wild animals as a hobby.

Few, if any, of the new zoos of the nineteenth and early twentieth centuries were started in ignorance. There was much which those already in being could teach about what needed to be done, and how to obtain animals. Following the lead of the London Society, some – for example, those of Philadelphia and New York and Antwerp – specifically declared that the zoos they were about to launch were to be concerned with science and education. When the London Zoological Society started in 1826, it immediately arranged scientific meetings at fortnightly intervals. It also launched a series of what soon became international scientific publications. A few of the newer zoos tried to do the same, but in most cases only lip-service was paid to what many regarded as secondary and unrealizable functions of a zoo. Some zoos, probably the majority, did not even have veterinary help to call upon on a regular basis. For example, the American National Zoo in Washington, founded in 1890, did not appoint a resident veterinary officer until 1941, and only in 1970 was it provided with up-to-date quarters for its scientific and veterinary work.

It is fairly obvious why Paris and London succeeded scientifically where other zoos failed. However shabby the zoo in the Jardin des Plantes may have been until recent years, it was part of an institution which remained in the mainstream not only of French but of international zoological enquiry certainly up to the time of the First World War in 1914. The situation in London was the same. From its very beginning, a large proportion, usually at least half, of the members of the Council of the Zoological Society of London were leaders in zoological science, while the Zoo itself, situated as it is in a royal park, has always had to live up to the high standards expected of a national institution which graces the capital of a country that, until the Second World War, was regarded as the dominant power in the world.

But there was another and related reason why the London and Paris zoos had a head-start as scientific institutions over zoos in other cities. As the capitals of their countries, Paris and London were the homes of many other scientific institutions. The Royal and the Linnean Societies were in London. The city was about to found its first university colleges, and Oxford and Cambridge were not far distant. Regent's Park was ringed by several major teaching hospitals, and for nearly a century the Society's meeting rooms, which were then no more than a mile or so from the Zoo in Regent's Park (today they are on the same site) were accessible to all interested scientists. In the first thirty years of its existence, the Society also managed a museum – consisting largely of specimens that came from the Zoo's post-mortem rooms – which was then more important than the zoological section of the British Museum. (The present Natural History Museum did not open its doors until 1881.) Small wonder that the Society's meetings were lively affairs from the start, and that the pages of its *Proceedings* and *Transactions*, launched in 1830 and 1833 respectively, were filled with major contributions by British zoologists. The London Society also straightaway established a category of Foreign Fellows, and those elected were the outstanding biologists of their time – men such as von Humboldt, Etienne Geoffroy Saint-Hilaire, Majendie, Blumenbach, Frédéric Cuvier, Temminck, Agassiz, Henry and Alphonse Milne-Edwards, Kölliker and Haeckel. Their names reflect the history of biology. The Presidents of related scientific societies, headed by the Royal Society, were *ex officio* honorary members. There was also a category of Corresponding Members, most of them Englishmen who were living in distant parts of the Empire, and who were anxious to further the purposes of the Society.

Many of the distinguished band of nineteenth-century Foreign Fellows, for example Cuvier and Agassiz, contributed to the Society's publications. There is a fascinating note from Etienne Geoffroy Saint-Hilaire, which was read at a meeting of the Society in February 1833.[8] In it he enquired about certain points that had been made the year before in a paper on the milk glands of the echidna, the primitive egg-laying mammal. The author was Richard Owen, who was to become the outstanding nineteenth-century figure in the study of fossils. Not convinced by Owen's observations, Saint-Hilaire sought further information. But, as he said, if the facts contained in Owen's article, partly based as they were on a communication made by a Dr Weatherhead to the Society's Committee of Science and Correspondence, 'should make me change my opinion, so much the better: I would rather be put right, than indulge in any views formed *a priori;* in this way I learn more; and it is for me always more gratifying to get rid of an error in science than to introduce into it an additional observation'. This unequivocal declaration of an essential scientific precept was remarkably forthright for its period, and illuminated, as much as it ever could be illuminated, the essential scientific purpose of the new Zoological Society of London.

Although it was also situated in a highly civilized and major capital city, the Paris Zoo was not as fortunate as was London's in its scientific relations, mainly because the museum with which it was associated was not as independent as the London Society. The Jardin du Roy had been established as a physic garden by Louis XIII in 1635, some ten years after the corresponding garden in Oxford, and forty years before the founding of the celebrated Chelsea Physic Garden in London. When his garden opened, Louis appointed his chief physician as superintendent and commanded that three doctors from the medical faculty of the University of Paris should also be attached to the garden in order to study medicinal plants and to start a 'cabinet', i.e. a collection of specimens. This was the humble start of the Muséum which, in 1793, in the opening years of the Revolution, became the Muséum d'Histoire Naturelle.[6] As one of its first acts, the new republican legislative assembly, which had replaced Louis XVI's absolute monarchy, decreed that the Muséum was to have a director under whom a number of professors were to run its separate departments. Before the middle of the eighteenth century, the King's cabinet of the Jardin du Roy already had 'chairs' of anthropology, palaeontology – the science of fossils – comparative anatomy and zoology. After the Revolution, and during a brief period in which all the universities of France were suppressed, the Muséum operated as the only scientific institution in the country, launching journals with which the Zoo could be said to have been associated.

On the other hand, the interest which the director and professors had in the Zoo varied considerably over the years, and it was not until 1917 that the responsibility for the menagerie and gardens was made the separate and full-time task of one man. This probably implied a partial separation of the Zoo from its more scientific base. But the base itself – that is to say the Muséum – was not a meeting-place for scientists in the same sense as was the Zoological Society of London. Public lectures were given in the Muséum, but the zoologists of France held their debates in the Institut des Sciences – a branch of the Académie Française – not in the Muséum or the Zoo.

The publication in 1858 of Charles Darwin's general theory of evolution by natural selection transformed the whole zoological scene, and certainly added a new dimension of interest to all existing zoos. But Darwin did not like the Zoological Society of London.[i] 'I am out of patience with the zoologists,' he wrote, 'not because they are overworked, but for their mean quarrelsome spirit. I went the other evening to the Zoological Society, where the speakers were snarling at each other in a manner anything but like that of gentlemen.' Of course, not all the Zoo's meetings were like that, and the record of the particular occasion which inspired Darwin's comment is hardly a picture of snarling zoologists. In view of his dislike of the Zoological Society, it is not surprising that his famous joint paper with Alfred Russel Wallace was presented to the Linnean Society and published in its *Journal*.

The subsequent debate about evolution, however, and particularly that part of it which relates to man's descent, made the London Zoological Society's meetings even more interesting than they had been before. Thomas Henry Huxley, the foremost exponent of Darwin's new theory, was a staunch supporter of the Society and a member of its Council. Philip Sclater, the Society's Secretary from 1859 to 1902, made his reputation through studies of the significance of the zoogeographical distribution of animals to the general theory of evolution. He, like later officials of the Society, was one of the foremost British zoologists of his day. Over the period from 1847 to 1942, when the post of Secretary to the Council and general administrator of the Zoo carried a salary, it was regarded by British zoologists not only as the most richly rewarded, but also as one of the most prestigious zoological appointments in the country.

It was only natural, therefore, that the London Zoo, with the country's leading professional zoologists always sharing the responsibility for its direction and management, should have been regarded throughout the world, and decade after decade, as a model of what a scientific zoo should be. With the nominal exceptions of the Jardin des Plantes in Paris and the Berlin Zoo, no other zoo in any country was the direct and continuing concern of its leading professional zoologists. And correspondingly, it is also not surprising that the list of the London Society's Foreign Fellows over the years includes very few directors of menageries – or 'zoo men' as they are now called – as opposed to zoologists, who were adding to the body of scientific knowledge.

The paramount position of London in the world of zoos was eloquently spelt out in 1929 at the Society's centenary celebrations.[20] On this occasion it fell to Sir Charles Sherrington – a scientist who stands in relation to our knowledge of the nervous system as William Harvey does to that of the circulation of the blood – to present a congratulatory address on behalf of both the Royal Society and of other British scientific societies. He referred to the 'long and honourable record of the Zoological Society in the promotion of research and exploration in one department of such knowledge, and in the education of a wider public in the results of its scientific activities'. And he then went on to say, 'To how many a child – and man and woman – has it not brought an unforgettable revelation of living Nature's variety and wealth. It has largessed knowledge to the public.' Similar sentiments were expressed by Dr

Gravier, speaking for the Académie des Sciences of Paris and the Muséum d'Histoire Naturelle, and who specifically referred to the heavy burden shouldered by the London Society in publishing, in addition to its *Proceedings* and *Transactions*, the *Zoological Record*, a work which is indispensable to all who are interested in the study of animals. (Today the Society also shoulders the responsibility for the *Nomenclator Zoologicus*, the official list of the names of genera and sub-genera in the animal kingdom, from the tenth edition of Linnaeus onwards, as well as the *International Zoo Yearbook* and a series of *Symposia* on specific zoological topics.) Dr Casey Wood, who represented the Smithsonian Institution of Washington, went even further than Dr Gravier:

> In our efforts to realize Smithson's wise programme we have found the experiences and plans of the London Zoological Society of pre-eminent value. Indeed, we may almost be said to have sincerely flattered you by a close imitation in many respects, although we can never hope to equal your unrivalled Gardens, mainly because we have not contributory agents in the shape of world-flung people who, truly devoted to the outdoor study of nature, bring or send all the treasures of animal life on this earth to your very doors.

THE FOURTH PHASE – A PERIOD OF DECAY

The years from 1918 to 1939, the interval between the two world wars, were a period of general economic and political upheaval. They can also be regarded as constituting a fourth phase in the evolution of zoos and, in general, one of scientific decay. Those zoos which had failed by the turn of the century to launch themselves scientifically and educationally continued as mere animal show-grounds. Sometimes rarities would crop up to add glamour, and on occasion there would even be a new species, such as the okapi, previously unknown to Europeans. Some zoos were able to improve their displays, and even a few new ones – such as Munich and San Diego – were started. The London Zoological Society opened a spacious country zoo at Whipsnade, where animals could live and breed in large paddocks and in conditions totally different from the inevitably confined quarters of a metropolitan zoo.

But during this period zoos started to pay less and less attention to new focal areas of zoological interest. With a few notable exceptions, for example the Philadelphia Zoo, they were not even adding anything very significant to the body of veterinary knowledge. Although there was a close link with the nearby Royal Veterinary College, the authorities of the London Zoo then regarded veterinarians as poor cousins of the medical profession. Indeed, the Council of the Society behaved as though it had no need for veterinary officers when it could always call on the best in human medicine. Eminent British physicians and surgeons were eager to study unusual clinical conditions in animals, and to use post-mortem material from the Zoo in their own researches. Routine studies of blood parasites and of worm infections in zoo animals were carried out by the two foremost authorities in the country. The Society also maintained the closest links with a major national laboratory which specialized in nutritional problems and vaccines. Moreover, it had its own qualified pathologist who was provided with adequate staff and facilities, while at least one of its permanent officials was always medically qualified. It was not until 1950 that the Council numbered amongst its members a professional veterinarian – but then he also happened to be the head of the Royal Veterinary College!

Scientifically the London Zoo was undoubtedly better off during the inter-war years than any other civic or national zoo – practically none of which had any meaningful scientific connections. Interest in comparative anatomy and in problems of human evolution remained powerful enough to fill the Society's meeting room and the pages of its journals. Nonetheless the Society was seemingly blind to the fact that it was moving into a scientific backwater. It was understandable that the national museum in Paris, like its counterpart the British Museum in London, should be overwhelmingly concerned with the identification of new specimens and their classification. But other than prejudice and conservatism, there was little reason why an institution such as the London Zoo, the foremost of its kind in the world, should have failed, in spite of all the scientific advantages which it enjoyed, to acknowledge the new course along which the mainstream of zoological science was going to flow. From the latter part of the nineteenth century, experiment had started to supplant anatomical study as the favoured way of extending zoological knowledge. The basic laws of inheritance which Mendel had enunciated had been re-discovered after they had been buried for several decades in the pages of an obscure journal. They represented the missing half of the concept of evolution, and the half which was open to disciplined scientific analysis by way of controlled experiment. To the bright younger zoologist, comparative anatomy and speculations about possible lines of evolutionary descent started to seem a dull way of illuminating the concept of evolution. Other fields of experimental enquiry, for example blood groups, immunology and the action of hormones, were also opening up.

In spite of all this ferment, the scientific zoo of London, like that of Paris, remained almost exclusively committed to the straightforward study of animal form and its classification. In 1902 Sclater had been succeeded, as Secretary of the Zoological Society of London and as chief executive of its Zoo, by Peter Chalmers Mitchell, a distinguished zoologist of the old school, and a man who was positively opposed to experimental work on animals. During the thirty years of Mitchell's secretaryship, no one on the staff of the Society was allowed to hold a licence under the British Vivisection Acts, even though this meant, to quote his own words,[1] a major limitation in 'the assistance which pathology could give us in the practical care of our animals'. When, in 1935, Julian Huxley, a scientist who had been one of the pioneers of experimental zoology in England, succeeded him as Secretary, he discovered that he too was bound by the same conservative convention. The only significant move he was able to make in the modernization of the Society's attitudes was to divide its published *Proceedings* into one series which dealt mainly with animal structure and classification, and a second to which he invited contributions from experimental biologists who, not being on the Society's staff, were not bound by the ruling about experiments on animals. His continuing battle with the more conservative members of the Society's Council led in 1942 to his resignation.

The fourth phase in the evolution of zoos can be fairly summed up as having been a period which was dominated all over the world by the forces of reaction, not just in the world of politics, but also in that of zoos. Zoological science was moving ahead fast, new societies to deal with particular fields of biological and zoological enquiry were being formed, but zoos remained stuck in an early mould. And the gulf between animal displays and 'science' was still sometimes being illustrated in very odd ways. There was one member of the Council of the Zoological Society of London who during the inter-war years still kept his own small menagerie, partly because he was fascinated by animals, but partly too because it pleased him to amaze his guests by putting members of his staff on ostriches and llamas and even sending them, thus mounted, to a meet of the local fox-hounds. There was another, also a wealthy land-owner, who had wild zebras broken to the shaft to pull a pony-trap. There were others whose concern could hardly have been described as scientific or educational.

THE FIFTH PHASE – CONSERVATION AND COMMERCE

The start of the Second World War can be regarded as the opening of the present phase in the evolution of zoos. During the five years of fighting most European zoos suffered severely, not only from bombs but also because there were neither men to look after wild animals, nor food with which to feed them. During this period many zoos of the North American continent also went downhill from lack of attention and resources. When peace came in 1945 there was no immediate let-up. In the post-war years food for people remained a far more pressing need than food for animals, and building materials were more urgently wanted for houses than for cages.

Nonetheless, and in spite of their dilapidated state, not only did interest in zoos persist, in places it even increased. During the years of German occupation, the zoos of Antwerp and Rotterdam consolidated and improved their positions as major civic centres. In the years which followed the end of hostilities, the same happened in Frankfurt. With few competing distractions, even the run-down London Zoo became a kind of Mecca for war-weary people, not only during the period of hostilities but also in the immediate post-war years. In 1946 it had nearly three million visitors. In 1950 the figure rose to 3,100,000, the highest in all the Zoo's history, mainly because of the attraction of a single newly-born polar bear cub. If not as marked, the same increased appeal of zoos also manifested itself in America.

The response of zoo managements to a growing public interest in animals varied both with national and local circumstances and, not surprisingly, with personal and political determination. The speed with which the inhabitants of East Berlin, now cut off from their old zoo, were provided with one of their own was truly remarkable. In Frankfurt-am-Main Dr Bernhard Grzimek, a veterinarian whom the American military authorities had appointed police chief, was achieving miracles in rebuilding the city zoo at a time when the Council of London's Zoological Society was still debating how its zoos were to be managed following the resignation in 1942 of its chief executive. As the first step in their post-war regeneration, certain zoos decided to embark upon major new buildings and exhibitions. London, in more conservative fashion, began its post-war reconstruction by building new hospital accommodation for animals and new service and staff quarters. In the implementation of a policy decision to steer the Society back to the mainstreams of biological science, these were then followed by the construction of two major research institutes, one on Comparative Medicine, the other on Comparative Physiology, both of which are still unique in the world. Cities which had never before even dreamt of having a zoo now wanted one and, building on the experience of others, set out to eclipse those that were long established.

The net effect of all this renewed public interest in zoos was two-fold. First, they became regarded as major attractions, as ventures which in many cities needed to be encouraged more than, say, opera houses or civic orchestras. Second, they were seen as money-spinners. The extent of the consequent multiplication of zoos in the post-war years is not precisely known, but some indication of the increase is given by the fact that the number listed in the *International Zoo Yearbook*[14] for 1959 was 309; the corresponding figure in the 1978 volume is 883. Whatever else, they have certainly more than doubled.

The multiplication of zoos was a heyday for animal dealers. Safari parks also started to mushroom as opportunist ventures in which dealers and circus proprietors joined forces with big land-owners for their mutual financial benefit. If we exclude businesses such as circuses and menageries associated with funfairs, it would be true to say that the history of zoos had taken a new turn. Exotic animals were now being exploited to make money which was going to be used neither for the advancement of knowledge nor for the well-being of the animals themselves. Wild beasts had become a kind of currency. They could help to keep roofs on country houses, and to buy mansions for dealers and animal impresarios.

But without any general awareness of what was happening, public attitudes to the exploitation of wild animals were also changing. By the end of the fifties, what had started years before as a movement to protect the wild fauna of the world became part of a much wider concern with environmental matters, of a general realization that the spread of human population and the growth of wealth was transforming, often deleteriously, a natural environment which should, if at all possible, be preserved for future generations.

There was, of course, nothing new in the understanding that cruelty to animals was a bad thing, and that for a variety of reasons the survival of many animal species was endangered. Long before conservation became an international concern, many countries had already taken action to protect wildlife. In Great Britain a Royal Society for the Prevention of Cruelty to Animals was founded as early as 1824. One to protect birds took shape in 1889, its members pledging themselves not to wear feathers or furs from any creature – other than the ostrich – that had not been killed for food. A Fauna Preservation Society, in whose establishment the Zoological Society of London played a major part, was set up in 1903, its original name being the Society for the Preservation of the Wild Fauna of the Empire. Conservation laws had been enacted in South Africa as early as 1858, and a vast reserve for wild animals had been set aside by the Transvaal government before the close of the century. Even though most of the founders of the New York Zoo were big-game hunters, they too were concerned that unrestricted hunting was endangering the wild fauna of North America.

Fairfield Osborn, then President of the New York Zoological Society, started a Conservation Foundation in 1948. Individual groups in many other countries had concerned themselves with problems of conservation. An international office, which is now part of the 1948 International Union for the Conservation of Nature (IUCN), was accordingly set up in 1928 in Brussels, under the direction of Professor J. M. Derscheid, to coordinate the work of all national movements in this field. In 1973 international agreement was reached for what is now called the Washington Convention about measures designed to regulate and control international trade in endangered species of animals and plants. With public feeling becoming increasingly vocal, many countries also introduced legislation to protect their own wildlife and to slow down commerce in wild animals, fur and feathers. Some also established national nature reserves, similar to, if smaller than, the Kruger National Park in South Africa and the Yellowstone Park in the United States. In 1961, a World Wildlife Fund was started in order to gain financial support for the effort to prevent the extermination of endangered species. Originally its purpose was to help the IUCN, but the Fund has now branched out on its own as an executive body.

But by the mid-1960s, what had begun as a sober concern about the extermination of wild animals had become part of a diffuse worldwide movement for the protection of the global environment whose future was seen to be threatened by population growth, the exhaustion of natural resources and pollution. And this, like most popular movements, soon sprouted a variety of enthusiastic but uncritical fringe groups whose aim was to protest against all manner of things – against industrialization, nuclear technology, food additives, chemical fertilizers. There were 'doom-watchers' proclaiming the end of mankind, and wildlife enthusiasts who had learnt from the television screen about animals in their native habitats and about 'nature in the raw', and who were not slow to protest that there should be an end to zoos. Those responsible for keeping wild animals in captivity, whether in zoos or in safari parks, soon found themselves constrained to justify publicly what they were doing.

The underlying sentiment of those who now campaign against the exploitation of wild animals is the same as it has always been – a deep-seated belief that it is wrong to do so, and that it is wrong to deny any of God's creatures their 'natural liberty'. It was 'in the name of humanity' that Seneca condemned the brutal animal spectacles of ancient Rome. When in 1789 the people of Paris rose against an absolute monarchy to demand liberty, equality and fraternity, they marched on

Versailles to insist that not only they but also the monarch's caged beasts should be set free.

Today a new brand of anti-zoo propagandist has entered the scene, with a message which disguises this basic and humane sentiment by declaring that it is not the principle of keeping captive animals which is wrong, but that zoos are badly run. To the best of my knowledge not one of these self-appointed zoo-reformers, not one of those who have written books or articles or who have spoken on radio or television about the wickedness of zoos, has either been involved in running a reputable zoo, or has the background and experience to understand what goes on in the research departments of a scientifically-run zoo.[11] Of course there is no zoo that could not be improved, but one cannot expect people who have not been responsible for the management of serious zoos to say how they should be run. One would no more expect this than call on Christian Scientists to inform the medical profession about the best techniques which should be employed in, say, open-heart surgery or the treatment of kidney failure.

Indeed it could well be asked whether, in their widely-publicized pronouncements, the amateur zoo-reformers do less good than harm in the achievement of a purpose that is wholly estimable. They almost always attack on too broad a front, firing indiscriminately, and often erratically, at good and bad zoos alike. So far as the former are concerned the message is directed not only at organizations which do not need to be persuaded, but at officials who know from hard experience how much, or how little, can be achieved in zoos with what resources may be available, and what the constraints are on possible alternatives of action. There is little satisfaction for them in knowing that uninformed and undeserved criticism is always retracted when challenged. But because the good zoos are attacked, the bad ones can lean back and do nothing, secure in the knowledge that the good zoos are in the same target area as themselves.

There is, as I have said, no zoo that could not be improved given the will and the necessary resources and, above all, given more knowledge than is yet available about the housing, the nutritional requirements, and the diseases and breeding of wild animals. Every responsible zoo, all that are described in this book and all others like them, recognize that whatever resources can be made available are needed in order to improve our knowledge of these matters. The bad zoo is as much the product of ignorance as of commercial greed or of any other consideration.

Obviously these views would not necessarily be shared by the commercial enterprises, circus proprietors and animal dealers who have spawned the safari parks and commercial animal collections that have emerged during the past twenty years. Their main purpose is to make profits that can be creamed off for enterprises which have nothing whatever to do with the welfare of animals, and they do this by catering for a public that wants the excitement of driving a car through paddocks of lions, giraffes and baboons. Those parks that cannot make money close. Those that can, continue. Although safari and so-called game parks are frequently in the news because of accidents to visitors or keepers; because of their untrained keepers and inadequate veterinary care; because of animals that escape; and because of allegations of wilful cruelty, there is always a public to whom they appeal. In a recent book on British zoos the author, Anthony Smith, writes that he ran into difficulties in obtaining information about what goes on both in zoos and safari parks, and he accuses 'zoos and zoo-men' of 'dishonesty, hypocrisy, in-fighting, secrecy and mere illegality,' quixotically adding, however, that he is 'entirely' on the side of those who manifest these qualities.[23] Unfortunately he lumps together with the commercial people who make money out of safari parks, professional men and women who would be totally opposed to the practices which he fails to condemn. There is no secrecy and no dishonesty in the management of zoos which are part of charitable organizations whose aim is the advance and spread of biological knowledge. Smith, like some others, fails to underline this as a real issue and one which firmly separates two lines of development that now mark the evolution of zoos.

Zoos exist because the people who want them to do so vastly outnumber those who would like to see them suppressed. Obviously public money, either from central government or local authority, would not be provided to keep zoos going if the situation were otherwise. But civic pride in a well-run zoo park, however magnificent it may be, is not enough. If those who are responsible for managing zoos do not agree with the minority who are opposed to zoos, they at least have to show that they are just as humane in their attitude to wild animals. They have to be able to demonstrate, and demonstrate publicly, that they are trying to abide by the standards that are set by the best zoos in their treatment of exotic animals – standards of the kind described in books such as those of Lee Crandall[4] of the New York Zoo or Heini Hediger[10] of Basle and Zürich. They have to show that they treat their public educational programmes seriously. Their keeper staff has to be properly trained. They should be able to demonstrate – not to the world of amateurs who write about zoos, but to the scientific world – that they are seriously concerned with research of the quality and kind which yields knowledge that is not only relevant to the health and breeding of wild animals, but also to man's own well-being. And finally; they should be able to show that they have a contribution to make to the preservation of species which have either become extinct in the wild or which are now endangered.[9,25] There can never be any limit to the amount of money the fulfilment of these objectives demand.

There is no publicly available information which suggests that any commercial zoo or safari park has contributed to any of these goals other than by claiming that in addition to making profits it helps in conservation. Paradoxically, the worldwide rash of safari parks was to some extent encouraged by the very protection which the newly formed African game reserves afforded wild animals. When animals are shielded from the natural processes that keep numbers in check, they inevitably multiply at an unacceptable rate. And inevitably, too, they begin to destroy their enclosed habitats, however extensive they may be. For this reason controlled hunting has always had to be allowed in, for example, the Kruger National Park in South Africa, where every year about eight hundred elephants are culled so that such tree cover as there is in an area bigger than Wales is not destroyed. When the move to save the white rhinoceros from extinction was started, it was said that no more than twenty still existed in the wild. They have since multiplied at such a rate that today there is a considerable commercial traffic in these animals. Lions, giraffes, various kind of antelope, baboons – these are the animals commonly seen in safari parks. They are there partly because protection in their native haunts has led to over-population and, as the World Wildlife Fund has discovered, partly because some African countries are all too ready to call in animal dealers to help them earn foreign currency.

Technical information between zoos such as those described in the chapters which follow has always been exchanged directly, freely and openly. There are also more formal exchanges through the International Union of Directors of Zoological Gardens, formed in 1935, and a variety of national organizations, such as the American Association of Zoological Parks and Aquariums, which was founded in 1924. The *International Zoo Yearbook*, which the Zoological Society of London launched and which has appeared annually since 1960, has made the field of exchange much wider than it has ever been. In the introduction to the first number of the *Yearbook* it was emphasized that it is in the interests of everybody to reconcile the need to conserve wildlife with the move to develop zoos, and that the aim of the new annual was to help make good the lack of published information about the progress being made in animal husbandry and management, about the specialized building requirements of animals and about the best methods of display.

Unfortunately all these voluntary and open exchanges have not proved sufficient to assure that the best practicable standards are consciously followed in all zoos. Sadly, too, no measures have yet been devised whereby this can be done. The obvious way would be through legislation. South Africa has laws which control the establishment of zoos. Holland has. Sweden has. Police regulations apply to zoos in the Federal Republic of Germany. But while several other countries have thought of following the same line, most hesitate to do so, essentially because of the opposition from the commercial zoos. For example, while there are state laws which supposedly control the establishment of zoos, and while there is a Federal Animal Welfare Act that is designed to ensure proper husbandry of zoo animals, the federal government of the USA has failed on several occasions to secure enough congressional support for the passage of a national bill which would place zoos under sensible control. Powerful interests are concerned to prevent this happening, for the good reason that the openness which marks the affairs of such zoos as those of London, New York, Antwerp or Philadelphia does not accord with the procedures of most commercial enterprises which make profits out of exotic animals.

This fundamental divergence of interest is well illustrated by what has happened recently in the United Kingdom. Because of its awareness that certain sections of the public were becoming concerned about the poor conditions of a number of the new zoos, the Zoological Society of London took the lead in 1966 in the establishment of an organization called the Federation of Zoological Gardens of Great Britain and Ireland. This Federation included not only zoos which were either charitable or civic undertakings, but also a few well-respected commercial collections, and its purpose was to define and agree professional standards which all zoos should try to meet on a voluntary basis, members of the Federation undertaking to submit themselves to regular inspection by a properly qualified team. Since it was believed that there was a chance that national legislation might be introduced to impose standards for zoos, it was thought useful if the zoos themselves showed what criteria could sensibly be set for such matters as housing, nutrition, veterinary care and security, and which could be made the basis of legal standards.

Not long after the formation of the Federation a private bill to control zoos was introduced into the House of Lords of the British parliament. After various amendments, including some made at the suggestion of the ministry which would have been charged with the implementation of legislation, given that any were enacted, the bill was passed without dissent, and then submitted to the House of Commons. At that point, and in order to oppose any form of legislation, the new commercial zoos and safari parks organized themselves in what they called the National Zoological Association (a strange title in view of the fact that the people concerned had nothing whatever to do with the science of zoology – any more than self-styled political 'ecologists' have to do with the science of ecology[8]). They succeeded in their opposition. Because of a parliamentary technicality, the proposed bill lapsed. A new but similar bill was then introduced in the House of Lords, where once more it passed all its stages. The general election of 1974 then stopped further progress. The bill was then passed a third time through the House of Lords. Once again it failed in the House of Commons. And there the matter still rests.

What the commercial zoos and their supporters found objectionable were demands of the mildest kind, but demands which would have meant an end to the secrecy which surrounds the operation of commercial zoos. The proposed bill stated that:

> It shall be the duty of the owner of a zoological garden to make and maintain proper provision for ensuring at that zoological garden –
>
> (a) the health and welfare of the animals in the collection;
> (b) the provision of adequate veterinary care;
> (c) that an adequate number of keepers are available at all times;
> (d) the safety of the visiting public and of the staff;
> (e) that the zoological garden is maintained in a proper state of repair;
> (f) that proper records are kept of births and acquisitions and of deaths and disposals of animals.

This was too much for the opposition! To this day, no one other than the commercial enterprises themselves knows what manner of veterinary or other care is provided for the animals which they exhibit to the public, or what standards of training they require of their staff. Whereas anyone can be informed about, say, the new arrivals or causes of death in a 'society' zoo, people who would wish for such information from all but the exceptional safari park would remain unsatisfied. They would indeed be lucky to be told how many visitors pass through a safari park's gates.

Government departments are rarely keen on adding to their administrative burdens, particularly when these cost money. But in the United Kingdom we have the anomaly that, because of public reaction against private individuals keeping dangerous animals in their backyards, a Dangerous Wild Animals Act came into force in 1976. As a result, no one can now keep in a garden shed a lion or a leopard, a bear or an ape unless he has been granted a licence, the cost of which may be £100 a year. Yet zoos and circuses are exempt from this act! Totally unqualified people can now keep dozens of lions or bears or giraffes in parks which may be less secure than what would be demanded of a private individual keeping one cheetah! And if anyone, whether keeper or visitor, were to get hurt in a safari park, the only redress would be by appeal to the Common Law. If legislation continues to be blocked by individuals who seem less keen on the welfare of wild animals than on that of those who wish to exploit them, one way round the difficulty might be to amend the Animals Act of 1976 by making it apply to people who wish to keep and display, not one or two, but numbers of animals, and by incorporating into it the minimum standards which were set out in the bill that was blocked in the House of Commons – and by raising the annual licence fee by an amount adequate to pay for proper inspection services.

But this will not be enough if zoos are to continue. Standards are one thing; justification for keeping exotic animals is another. As I have said, zoos continue to exist because far more people want them than wish to see them suppressed. But it is not enough if, as Dr Grzimek[12] has suggested, only one in a hundred of those who visit zoos do so because they want to learn about wild animals, as opposed to the ninety-nine who do so for other reasons. If zoos are to continue, they must show that, in addition to providing recreation, they are also properly used to provide organized public education about animals. Above all, they should be able to show that the wild animals which are kept in captivity are being studied with the best resources of modern science to provide information not only of benefit to their own survival but also to that of man himself.

PARIS

Ménagerie du Jardin des Plantes
and
Parc Zoologique de Paris
du Bois de Vincennes

Marcia Edwards

No zoo has had so colourful a history as that of the Ménagerie du Jardin des Plantes in Paris; and few have been associated with so many of the great names of classical zoology. The story begins in the mid-seventeenth century, in the reign of Louis XIV, a monarch to whom the world owes the splendours of Versailles. When the King started to embellish his vast palace to the west of Paris, he decided to include in the gardens a menagerie consisting, not of isolated houses, but of a series of cages set in a coherent pattern within a splendid garden. The first animals were installed in 1665, and the collection then grew, mainly as a result of gifts from foreign royalty and heads of state. Initially, the privilege of visiting the King's beasts was restricted, but gradually more and more people were admitted, until the damage caused by unwelcome crowds thronging his park induced the King to deny entry to any but members of his court.

Both science and art gained from the establishment of the menagerie. Members of the Académie des Sciences were able to dissect the animals which died, and their studies helped to prepare the ground for the flowering of comparative anatomy in the following century. The King commissioned paintings and sculptures of new creatures as they arrived and many artists worked at Versailles during the last part of the seventeenth century.

Towards the end of his life Louis seldom visited Versailles. He also took less and less interest in his menagerie, but until his death in 1715 provision was made for its inmates. His successor, Louis XV, did not share his great-grandfather's interest in animals, and when his grandson Louis XVI became king in 1774, the menagerie, which was already more than one hundred years old, was in serious need of repair. The new monarch was equally unconcerned and it was not until 1785 that some much-needed restoration was undertaken. By then, however, revolution was stirring and in October 1789 Louis abandoned Versailles for Paris.

Not surprisingly, economies had to be made in the royal budget, and the collection at Versailles had to be reduced. But worse was to come. For some time menageries had been viewed with disfavour by many who held that it was shameful to feed animals when people were starving. In August 1792 a local group sympathetic to the Jacobins, the most ruthless element of the Revolutionaries, made its way to the menagerie. The leader told Laimant, the menagerie director, that they had come, both in the name of the people and of nature, to demand that those creatures which God had created to be free, and whose freedom had been denied by the vainglory of tyrants, should be set at liberty.

Laimant was in no position to refuse but, handing over the keys, he pointed out that the first act of some of the animals might well be to attack their liberators. Not unnaturally, this produced a change of plan, and it was decided that, for the time being, the more dangerous beasts should be left where they were. Harmless animals were released but, with a Gallic practicality which owed nothing 'to the name of nature' and even less to 'freedom', most were delivered to the butcher.

The Registrar-General for the area then wrote to Bernadin de Saint-Pierre, Comptroller of the Jardin du Roy in Paris, to enquire whether he wanted any of the few remaining animals, adding as an incentive that the collection still contained a superb rhinoceros. On his return from a visit of inspection to Versailles, Saint-Pierre recommended that all the animals should be brought to Paris to form the nucleus of a national menagerie.

The Jardin du Roy or, more fully, the Jardin Royal des Herbes Médicinales, had been proposed in 1626 by Louis XIII for the study of plants and their medicinal properties. Over the years the activities had widened to include other branches of natural history and medicine, each with its own chair and professor. About the middle of the eighteenth century, Buffon, the celebrated naturalist, developed and improved the Garden. He built conservatories and made a pond for waterfowl, and he also allowed peacocks to roam the grounds. A plan which he commissioned in 1776 provided for a great aviary and enclosures for other animals but, although he succeeded in enlarging the area of the Garden, he did not pursue the idea of a menagerie. That came into being almost by accident while negotiations were still proceeding to save the survivors of the Versailles collection.

In 1793 the Jardin du Roy became part of the newly established Muséum National d'Histoire Naturelle. Saint-Pierre had foreseen how the prospective museum and a menagerie could cooperate in the interests of public education. What brought his ideas to fruition was the fact that, at that time, Paris was the focus of travelling side-shows in which combats were staged between various sorts of domestic and wild animals. The civic authorities decided to ban these cruel contests and the police confiscated the animals, which were then foisted on the Jardin du Roy, by then known as the Jardin des Plantes. The Officers and Professors of the Jardin were, not surprisingly, dismayed, and some tried to reject the unwanted gifts. No preparations had been made to receive the animals but, equally, it was impossible to slaughter and dissect so many creatures all at once. As if these were not problems enough, the Garden was besieged by angry owners demanding compensation for the loss of their stock-in-trade. In the end shelter was found in some not particularly suitable buildings, and the question of care was temporarily solved by employing the animals' erstwhile owners. The national menagerie was thus founded with a small collection of creatures rescued from an unenviable fate. A few months later, in April 1794, the survivors of the Versailles menagerie, including a quagga, and a lion with his friend, a small dog, were added to the collection. The rhinoceros, ill from a wound received as the result of an accident, could not be moved and died at Versailles in May. That same month, a number of exotic animals from the estate of the Duc d'Orléans joined those in the Jardin des Plantes.

Better, though still temporary, quarters for the animals were soon built, and the Professors planned to establish a menagerie which would be of educational value to the nation, a project which had the sympathy of the Convention by which Paris was ruled. But times were becoming increasingly hard and, without regular funds to provide for the animals, many of them died of starvation and disease.

The situation took a turn for the better at the end of the following year when Mordant-Delaunay, who as 'citizen Launay' had recently been appointed to the staff of the Muséum Library, took matters in hand. Launay, a cultivated man with some literary talent, did not lack initiative. He brought together a collection of native French animals and, by various means which included as 'spoils of war' a valuable collection of animals from the Château of Het Loo, abandoned when the French army invaded Holland, improved and enlarged the menagerie. Additional land along the Seine was bought for the growing collection, and he also managed to arrange a proper budget for the menagerie. But within four years, when 'troubles' once again broke out, animals starved as food supplies dwindled. Not until the restoration of peace at the turn of the century were plans put in hand to restore and improve the Jardin. The architect Molinos designed a Grand Rotunda in the form of the Cross of the Legion of Honour. While this might have been a timely tribute to Napoleon, who was much interested in the menagerie and museum as he was in educational and cultural matters generally, it was not the most practical of ideas for housing animals. The building was first intended as a lion house, for which it was clearly unsuitable, and construction work was suspended while it was redesigned to house large herbivores. The year of its completion, 1812, was, alas, the year which saw the commencement of the decline of the man it was intended to glorify. Some rustic-style animal houses had been built in the preceding years and, in 1804. Frédéric Cuvier, brother of the great anatomist Georges Cuvier and himself the first incumbent of a chair of general physiology created in 1837, was appointed Director. He not only expanded the building programme, but acquired more land. Between 1817 and 1837 he added a Lion House, a Birds of Prey Aviary, a Pheasantry and a Monkey House, although in 1814 he had to suffer the anxiety of seeing Prussian troops about to take up quarters in the Garden. Fortunately the high reputation of the museum and its menagerie saved them from damage. The famous German naturalist Alexander von Humboldt had intervened on their behalf.

The middle years of the nineteenth century were uneventful for the museum and its menagerie. The collection was enriched as a result of increasing foreign travel and expeditions, but no 'new arrival' was to stimulate as much interest as did a giraffe presented as a gift in 1826 by the Pasha of Egypt. The young animal was put on board ship at Alexandria, together with cows to provide her with milk, four Arabs to care for her, and a parchment 'necklace' inscribed with passages from the Koran to ward off evil. She was offloaded at Marseilles in November, and the following year made her way, on foot and accompanied by her court, to Paris. When the entourage reached Saint-Cloud in June 1827, the King, Charles x, fed the giraffe rose petals from his own hand. Amidst great public excitement, she eventually arrived at the menagerie. The first of her kind to be seen alive in France, she was celebrated in song, prose and even fashion.

Anyone travelling abroad and willing to collect animals was pressed into service by Constant Duméril, a professor at the Muséum who was particularly interested in amphibians and reptiles. His growing collection of reptiles, amphibians, fishes and insects was accommodated in the old Monkey House and became the prototype of similar collections in other European zoos.

Much of the early work of the menagerie concerned itself with the acclimatization of exotic species and the hybridization of domestic animals, an interest which reflected the fact that its establishment, during the difficult time of the Revolution, had been partly justified by the belief that the zoo could produce additional sources of food for the populace. Isidore Geoffroy Saint-Hilaire, another distinguished zoologist of the period, was much interested in this aspect of the menagerie's work, and in the late 1850s helped to found a society whose purpose was to try to add to the number of domesticated animals. As part of this venture, which was paralleled by the formation of a similar society in England, a Jardin d'Acclimatation was established in the Bois de Boulogne.

The Franco-Prussian war broke out in 1870 and by late summer it became apparent that the invading Prussian army would soon be in Paris. The staff of the Jardin d'Acclimatation immediately decided to disperse their animals and many were sent to safety. But before all of them could be moved, military operations put an end to railway transport. Alphonse Milne-Edwards, yet another famous name in the annals of zoology, then offered to house the stranded animals, among them two elephants, Castor and Pollux, in the Jardin des Plantes. In the autumn the Prussian army laid siege to Paris, and by the winter famine in the city had become so acute that all edible animals in the Jardin des Plantes, including those from the Jardin d'Acclimatation, were slaughtered to provide food for the people, just as, not a hundred years before, the menagerie at Versailles had been decimated for the same purpose.

This was not the only calamity the Garden suffered. A field hospital had been set up in the grounds and on the night of 8 January 1871 the Prussians began to bombard the whole of the area. The shelling caused extensive damage, not only to the Muséum and menagerie but also to the hospital nearby.

The war ended in 1871 and, once again, a rebuilding programme had to be instituted. An imposing Reptile House, containing two large halls – one for crocodilians and the other for an aquarium – was put up in 1874. Other reptiles exhibited included lizards, snakes, turtles and tortoises. In 1881 another Pheasantry was completed, and in 1888 a large aviary, enclosing a small pond and many full-grown trees, was erected. This aviary was generally held to be the most elegant and advanced of its kind anywhere.

The appearance of the Jardin des Plantes then remained more or less unchanged until the 1920s. A Vivarium was built in 1926 for delicate animals needing special conditions of warmth and humidity. Still in use today, it houses insects and other invertebrates as well as amphibians, reptiles and various small mammals.

Several buildings dating from Cuvier's time had to be demolished. The old Lion House was replaced in 1937, and the new building now accommodates a representative collection of the cat family. Lions, leopards and tigers live there all the year round, but during the summer months the smaller cats are housed, with various canine species, in a Small Carnivore House.

The Monkey House, which dated from the mid-1830s, was pulled down in 1927 and, after some years in temporary quarters, the monkeys and apes were moved to a new home in 1934. Here baboons, drills and mangabeys breed well. The gibbons were provided with a large, high cage in which they have ample scope for swinging as they do in the wild. Many other outmoded buildings were also removed during the 1930s. Bars were replaced with modern barriers and the whole Garden took on a more pleasing aspect.

It was during this period of redevelopment that the sister institution of the Jardin des Plantes, the Parc Zoologique de Paris du Bois de Vincennes, was founded. Some seventy years earlier the Muséum had been granted land in the Vincennes woods on the outskirts of Paris, but had not developed the area. In 1932 it was exchanged for a more suitable site, where the City and Muséum agreed to establish a zoological park. The Muséum purchased the animals from a small zoo which had been part of the International Colonial Exhibition in Paris in 1931, and the new park, 43 acres (17 hectares) in extent, was opened in June 1934. It was designed by Charles Letrosne after the style of Hagenbeck, the famous animal dealer and trainer who, in 1907, opened the Stellingen Animal Park near Hamburg. Hagenbeck pioneered the concept of keeping animals in moated enclosures without bars and against backgrounds of artificial rocks. The use of vegetation and of hidden moats created dramatic panoramas in which various animals, both predator and prey, appeared to be living together.

Vincennes Zoo provides spectacular settings for large groups of animals. Spacious enclosures are landscaped to simulate various natural habitats and are separated from the public by moats or ditches. In many of the enclosures, artificial rocks serve to disguise the large indoor quarters which in winter are open to the public,

Where practical, suitable numbers of animals are kept together in breeding groups. There are nearly 700 birds belonging to about 140 species, and breeding colonies of flamingos are of particular interest. The Great Aviary houses birds of prey, and several species of penguin can be seen in the pools round the base of the Grand Rocher, an artificial rock rising to a height of nearly 240 feet (73 metres). Set in the rock are air-conditioned temperature-controlled caves into which the penguins can retreat when the weather becomes too warm. The Grand Rocher, with trails winding around it, also provides a suitable home for mountain goats and sheep, while inside a lift carries visitors to the summit for a bird's-eye view of Paris.

The Parc's collections of hoofed animals and bears are justifiably famous. About forty species of antelope are kept; most varieties of zebra and many kinds of deer, including the rare Eld's deer, are represented. There have been over fifty giraffe births, and the seven specimens of okapi were also born in the Parc. Okapi are found only in forested areas of the Congo, and are rarely seen in captivity. There are ten species of bear, including the South American spectacled bear, the North American grizzly and black bears, the kodiak from Alaska, polar bears, and bears from India and Malaysia. The great apes and monkeys are well represented. Some live in semi-liberty on islands in the Parc, and baboons in particular have bred well for many years.

In 1973, to mark President Pompidou's visit to Peking, the People's Republic of China presented two giant pandas to the people of France. One did not survive long but the young male, now installed at the Parc, remains in good health.

A good collection of ungulates, which complements that at Vincennes – both institutions being directed by Professor Jacques Nouvel – is kept in the Jardin des Plantes, and its breeding successes include the Przewalski horse and the European bison. Among the *Bovidae*, gorals, bharals, several types of ibex and a breeding group of chamois are of particular interest. There are also deer, and the domesticated llamas and alpacas.

Many of the early menagerie buildings in the Jardin des Plantes are now protected as national monuments, including the rustic-style houses built in the early nineteenth century and now used to house some of the ungulates. The Rotunda, built to commemorate Napoleon, accommodates camels and a selection of mammals and aquatic birds which need shelter in bad weather. Its central dome contains an aviary for tropical birds.

Other houses which have been renovated include the Old Pheasantry, now used for the hardier doves and pheasants, the New Pheasantry, designed for birds needing warm winter quarters, and the Great Aviary, in which aquatic birds breed successfully. The Birds of Prey Aviary is particularly noteworthy for its display of European and African eagles, and many birds wander freely in the grounds of the menagerie as their predecessors did in Buffon's day. The Bear Pit, dating from the foundation of the menagerie, has been restored, and houses European brown bears. A new enclosure for wolves was completed in the early 1960s. Several of the world's rare reptiles, including an important collection from Madagascar, are on view in the Reptile House.

From the moment, nearly two hundred years ago, that it came under the control of the Muséum National d'Histoire Naturelle, the menagerie of the Jardin des Plantes became involved in scientific research. The first Director to be given the title of Professor was, as already mentioned, Frédéric Cuvier. The Director who first had joint charge of the Jardin des Plantes in Paris and the Zoo at Vincennes was Achille Urbain, a scientist particularly interested in animal nutrition. In his inaugural lecture 'Leçon inaugurale du Cours d'Ethologie des animaux Sauvages' (*Archs Mus. natn. Hist. nat.* Paris (6) 12 : 1–79), he outlined a programme of research into the behaviour, diets and disorders of animals, including those diseases which can be transmitted to man. Since his day, research at the two collections has expanded in scope and variety but it still concentrates on the areas he denoted, in particular parasitology.

In the seventeenth century the Académie des Sciences had put the King's menagerie at Versailles to good use and, in so doing, prevented it from being merely another spectacular collection assembled for no better purpose than to reflect glory on its owner. In the same way, the close connection of the staff of the Muséum and that of the two animal collections has been of inestimable value in the advancement of scientific knowledge. Lacépède, Cuvier, Valenciennes, Latreille, Lamarck and other great names in the history of zoology all held office at the Muséum.

The *Annales du Muséum National d'Histoire Naturelle*, now known as the *Archives*, have appeared without interruption since 1802. The Muséum is also responsible for the *Bulletin*, *Memoirs* and nearly forty other specialized journals. Its central library incorporated that of the Jardin du Roy and, although primarily a research library, it contains many rare books and treasures confiscated during the Revolution from the collections of *émigrés* and of religious houses.

Learning on a somewhat lower academic level has not been neglected. Documentary films on animal biology are produced and at both the Ménagerie and the Parc there are specially arranged conducted tours for schoolchildren.

Three and a half centuries have seen one king's small physic garden develop into a world-renowned museum, and another king's menagerie transformed into the national collection of living animals. 'Politics', it has been averred, 'is the art of the possible.' At times it must have seemed to the officers of the Jardin des Plantes that the politicians were bent on proving the opposite. But they need not have worried. Today the Jardin, the Parc and the State combine to guard and develop an incomparable heritage.

LONDON

The London Zoo
and
Whipsnade Zoo

Peter Olney

The London Zoo is Britain's national zoo, in the same sense that Paris Zoo is that of France, the Moscow Zoo that of Russia and the Washington Zoo that of the United States. But it differs from all other national zoos and, indeed, from most civic zoos, in one major respect. It is not supported by the State, except in so far as the land it occupies is Crown land for which no rent is charged. Although a national institution, the Zoo is run as an educational charity under Royal Charter.

The Zoological Society of London, the custodian of the Zoo, was founded in 1826, towards the end of a movement which began in the latter part of the eighteenth century when specialized scientific societies, commencing with the formation of the Linnean Society in 1788, started to take over areas of science which up to then had been under the umbrella of the Royal Society of London. The sponsorship exercised by the Zoological Society, the premier society of its kind in the world, is what gives the Zoo its unique character – that of having been managed from the start on a proper scientific basis.

The Society came into being because of a widely expressed need for an organization that could provide opportunities for the development of zoology as a separate subject. In 1817 Sir Stamford Raffles, the British colonial administrator who is best remembered for having founded Singapore, suggested to Sir Joseph Banks, then President of the Royal Society, that a collection of animals should be established in London for the purposes of study and general interest. The idea was pursued with Sir Humphry Davy, Banks's successor who, with Raffles, was responsible for drawing up a prospectus for a Zoological Society of London that was issued in March 1825 and which stated:

> It has long been a matter of deep regret to the cultivators of Natural History that we possess no great scientific establishments either for teaching or elucidating zoology, and no public menageries or collections of living animals where their nature, properties, and habits may be studied.... Should the Society flourish and succeed, it will not only be useful in common life, but would likewise promote the best and most extensive objects of the Scientific History of Animated Nature, and offer a collection of living animals such as never yet existed in ancient or modern times ... animals to be brought from every part of the globe to be applied either to some useful purpose, or as objects of scientific research, not of vulgar admiration.

In the letter he wrote in April 1825 to Sir Robert Inglis, an influential member of parliament and a Fellow of the Royal Society, Raffles emphasized that the Society should be primarily a scientific body. Raffles wrote: 'In the first instance we look mainly to the country gentleman for support, in point of numbers: but the character of the institution must of course depend on the proportion of men of science and sound principles which it contains. *I* look more to the scientific part, and purpose . . .'

There was obviously a need for such a society since some hundred people attended the first general meeting, held on 29 April 1826. Sir Stamford Raffles was elected President, and several resolutions for the formal establishment of the Society were adopted. The first one stated: 'That a Society to be designated the "Zoological Society" be instituted for the advancement of zoological knowledge,' and 'That the attention of the Society be directed to the following objects: the formation of a collection of living animals, a museum of preserved animals, with a collection of comparative anatomy; and a library connected with the subject.' Raffles did not live to see the fruits of his initiative; he died later that year of a stroke.

Raffles was succeeded as president by the third Marquis of Landsdowne, an eminent Liberal parliamentarian and a Fellow of the Royal Society. His term of office lasted only four years, but during his time and that of Lord Stanley after him, a great deal was achieved. Ground for the Zoological Gardens in Regent's Park was obtained from the Crown Estate at a peppercorn rent; buildings were erected and the area landscaped; a museum and offices were established in Bruton Street near Piccadilly; a farm was started in Richmond Park for rearing and experimenting with breeding, mainly in the interests of acclimatization (the farm was given up in 1834); monthly business meetings were begun and gradually assumed the character of the scientific meetings that are still held; a library was founded; and the Society's first scientific publications appeared.

In 1829 King George IV granted a Royal Charter 'constituting and founding a society . . . for the advancement of Zoology and Animal Physiology and the introduction of new and curious subjects of the Animal Kingdom, by the name of "The Zoological Society of London".' These have remained the aims of the Society ever since. Although its scientific and charitable status has been challenged (in 1854, 1930 and 1958 – see Zuckerman 1976), the courts decreed that the Society could not lose its scientific character without violating its charter – 'for science is at the very heart of our constitution'.

On 27 April 1828, the Gardens were opened to members and their guests. The Zoo proved to be so popular that in 1830 the number of visitors recorded was 224,745. There were fewer than two dozen staff in the offices and Gardens, while the list of members of January 1829 contained the names of 1,294 ordinary fellows and 40 honorary and corresponding members. The annual subscription for Fellows was first set at £2 and then, four years later, at £3. It remained at this level until

1958 when, despite strong protests from a group of Fellows, it was raised to £8. While an exact comparison is impossible, it is worth noting that in 1850 £3 was what an agricultural labourer earned on average in eight weeks – and there were seven working days in his week – whereas the present annual subscription of £10 represents about one day's basic agricultural wage for a five-day week. The restriction of admission to members only was not lifted until 1847, when the public was admitted on weekdays on payment of one shilling. Sundays were reserved for members only until 1940, when the privilege was reduced to Sunday mornings only, and in 1957 was withdrawn entirely.

The original site has been enlarged, and today the Gardens occupy 36 acres (14.5 hectares) in the north-east corner of Regent's Park. The area is divided by the Park's main circular road and by a stretch of the Regent's Canal into a Main Garden, a Middle Garden and a North Garden. In 1927, Sir Peter Chalmers Mitchell, the Secretary of the Society, organized the purchase of some derelict farmland 50 miles (80 kilometres) north of London. He and the Superintendent, Dr G. M. Vevers, transformed the area into parkland containing open grassy paddocks where animals could be kept in near-natural conditions, and Whipsnade, the prototype for all later zoological parks, was opened to the public in 1931. It was an immediate success, some 26,000 people visiting it on the first Whit Monday. The London Zoo and Whipsnade Park are complementary, and together house some 8,230 vertebrate specimens of about 1,250 species. The Society is still governed by an elected President, Treasurer, Secretary and a Council of eighteen Fellows, who have the ultimate responsibility for the management and conduct of the Society and for the appointment of staff. There are now nearly five hundred full-time permanent staff, including a high proportion of scientifically qualified personnel. Nearly two hundred of the staff are responsible for the care and management of the animals, while about the same number deal with amenities for the public. Other staff deal with administration and education. The Society possesses one of the most comprehensive zoological libraries in the world, two research laboratories, a hospital, a pathology department, a quarantine station and an education centre.

Landscaping and the design of buildings have always been subject to the agreement of the Society's landlords, the Crown Estate and, in later years, to the British Fine Arts Commission. Zoo architects, with the help and advice of the Society's officials and keepers, have designed houses for animals in the light of the best available knowledge about the husbanding of exotic animals.

The first architect of the Gardens was Decimus Burton who, although young, was already well-known in the 1820s. When only twenty-three he had designed a Coliseum in Regent's Park as an exhibition hall and place of public entertainment. It was built in the contemporary classical style with a dome larger than that of St Paul's. The venture proved a failure, however, and the site is now occupied by terraced houses. Two years later Burton designed the new layout of Hyde Park and the Triumphal Arch at Hyde Park Corner. He was also responsible for several other major buildings in London, including the Athenaeum, and the Palm House at Kew Gardens.

Burton's treatment of the first 7 acres (2.8 hectares) of Park to be leased to the Society was to influence the development of the Zoo for many years. From the start the Gardens had been a place in which to walk and, for many years, among the most fashionable pleasure-gardens in London. In 1828 the *Mirror* reported: 'The Gardens, independent of their zoological attractions, are a delightful promenade, being laid out with great taste, and the parterres boasting a beautiful display of flowers. The animals, too, are seen to much greater advantage than when shut up in a menagerie, and have the luxury of fresh air, instead of unwholesome respiration in a room or caravan.'

Among the first animals to be presented to the Society was a griffon vulture called Dr Brookes (it lived for another forty years), a white-headed eagle and a female deer from Sangor. Gifts were received from many parts of the world, especially from countries in the spreading colonial empire. By 1827, nearly two hundred animals were already housed in the menagerie, including 'two beautiful Llamas, a Leopard, Kangaroos, a Russian Bear, Ratel, Ichneumon, etc., etc., besides a pair of Emus, Cranes, Gulls, Gannets, Corvorants [*sic*], various gallinaceous birds and many others'. Other early acquisitions included elephant and quagga (now extinct) in 1831, Indian rhinoceros in 1834, giraffes in 1836 which, with the addition of two females in 1849, started a herd which survived until 1881, orang-utan in 1837, the first echidna to be seen alive in Europe in 1845, the hippopotamus Obaysch in 1850, giant anteater, the first to be seen alive in Europe, in 1853, Mallee fowl in 1854, Arabian oryx, not previously known to scientists, in 1856, giant salamander in 1859, kudu and steinbock, again the first to be seen alive in Europe, in 1861, aye-aye and lesser birds-of-paradise in 1862, the African elephant Jumbo in 1865, black rhinoceros in 1868, Père David's deer in 1869, anoa in 1871, Sumatran rhinoceros in 1872, and Aldabra giant tortoises in 1875.

Some of the earliest buildings were 'small follies set in an elegant garden'. Others, such as the curved-fronted aviary of about 1830, with its fluted columns in the shape of palm stems, were somewhat more functional. But in many cases the buildings appear to have been designed more for the pleasure they could give to visitors than for the needs of their inmates. Fragments of Burton's work survive. The cage which now sits empty on the Fellows' Lawn was constructed in 1829 to house a pair of king vultures, for which it was completely unsuitable. Later it housed macaws and ravens.

The Giraffe House, built in 1837 from a design by Burton, is now incorporated into the Cotton Terraces on the north bank. The original Llamas' Hut, which had a clock tower added to it ('the expense of the clock and building is not to exceed £100, including the expense of putting up the same in the turret' – minutes of a meeting of 18 March 1829), later housed camels and is now used for storing pushchairs!

Burton's East Tunnel, which runs under the public road separating the Main and Middle Gardens, remains basically unaltered and looks as pleasing now as it did when built in 1829. His plans for a graceful and practical carnivore house were never realized. At that time there was a difference of opinion as to whether most animals should have heated houses or be left in the open, and the advocates of heat triumphed. In 1843 the Carnivora Terraces, later known as the Bear Terraces, were built to a design by a lesser-known architect, Mr Elmslie, and although he favoured open-air display, his building was inferior to Burton's design, the dens being dark, small and draughty. Nevertheless these Terraces, with some modifications, were a notable feature of the Zoo up to 1919, and part of their public promenade remained until 1970. In 1849 the first-ever reptile house was built and, in 1850, the first-ever aquarium.

Apart from the Carnivora Terraces, no substantial buildings of any note were erected until in 1876 a neo-classical lion house was opened. This had a pleasing exterior and provided good facilities for handling and moving the animals, but because the indoor dens faced on to the public promenade the animals had little privacy. By 1877 outdoor enclosures had been added, but since they faced north the animals got little sunlight. In 1881 the Society opened the first insectarium in any zoo. It was a simple metal-framed glass greenhouse where the temperature was kept high and in which cases of insects, including such aquatic species as caddisfly and dragonfly larvae, were displayed. The Insectarium was replaced by the present Insect House in 1912.

The Zoo's main bird house of today started life as a reptile house which was opened in 1883. The use of cast-iron columns and a prefabricated wide-span structure gives plenty of room for visitors and birds, and still imparts a feeling of Victorian elegance. Already by 1887 there were signs of a more naturalistic approach to design and landscaping. In that year the Great Aviary was built, a beautifully simple enclosed structure, landscaped to provide natural surroundings for its inhabitants. Pairs of straw-necked and sacred ibis almost immediately nested and reared young. This highly successful aviary was in use until 1970, when it was dismantled to make way for the present Sobell Pavilions for Apes and Monkeys. The design of the Great Aviary probably influenced the design of the Southern Aviary and sealion pool and enclosure which were built in 1905, and which are still among the most functional and pleasing structures in the Zoo.

The next major building project was the Mappin Terraces, named after their donor and opened in 1913. Their design was influenced by the scenic panoramas used by Carl Hagenbeck to display animals in Hamburg, and by similar projects in Antwerp. The three-tiered Terraces, built of ferro-concrete to simulate natural rock, provide the maximum number of enclosures possible in a relatively small space. In 1924 an aquarium was built under the Terraces.

In 1927 the present Reptile House was opened, to replace the 1883 one. The new design owed much to the knowledge and advice of Miss Joan Proctor, the Curator of Reptiles at the time. In the 1930s, the most notable buildings to be constructed were those designed by Berthold Lubetkin and others in the Tecton group of architects. As innovators of many new architectural ideas, especially in the use of reinforced concrete, they were responsible for the Gorilla House (1933), Penguin Pool (1934) and, at Whipsnade, the Elephant House (1935). The Gorilla House and Penguin Pool are now listed under the Town and Country Planning Act of 1962 as being of particular architectural or historic interest, and cannot be altered. Although of undoubted merit, the buildings' statutory permanence imposes severe limitations on their present-day-use.

Little major development took place during the twenty years starting in 1935, when Sir Julian Huxley was Secretary. The arrival of Ming, the Zoo's first giant panda, probably helped boost gate numbers, but Huxley's proposed methods of modernizing and revitalizing the zoological image of the Society were not acceptable to all, and he resigned in 1942. In 1946, after the strain of the war years, the number of visitors reached an unprecedented 2,750,000, and in 1950 the attraction of the well-publicized baby polar bear, Brumas, helped push the numbers to over three million – a figure not repeated since.

The mid-1950s were the start of an era of great development, both in building and in the Society's scientific and educational activities. In 1958 an ambitious overall plan for transforming the existing Zoo was adopted. Although the Society in general was able to meet its operating costs, it lacked the capital needed for a comprehensive redevelopment programme. It managed to find some money from its own resources for the programme, but most of it came from foundations and from private benefactors. The large foundations, including Ford, Nuffield and Wellcome, helped to establish new research institutes. The Wolfson Foundation made it possible to modernize the Society's headquarters and library. Individuals, in particular Sir Charles Clore, Sir Michael Sobell and Mr Jack Cotton, donated large sums for specific building projects. Then, in 1970, the government acknowledged its share of responsibility. It was announced in parliament that a substantial measure of support would be provided because 'the Zoo has become in fact, if not in form, a national institution . . . it is a major London amenity, and also an important tourist attraction'. This mark of confidence, which was expressed in the form of a grant and in the waiving of the repayment of a government loan, made it possible for the Society to continue its redevelopment programme at Regent's Park and Whipsnade.

The first parts of the 1958 plan to be put into effect were the more obvious essentials: new service buildings and a boiler house and, in 1961, a new bridge joining the Middle and North Gardens across the Regent's Canal. In 1963 the new Cotton Terraces alongside the Canal were completed. They incorporate Burton's Giraffe House of 1837 in the middle of a terrace of simple houses and paddocks, and by clever landscaping and the use of moats give the viewer a feeling of space. The Terraces house mainly ungulates such as zebra, camel, giraffe, bison, deer and antelope. Opposite the Cotton Terraces, on the north side of the Canal, is the large Snowdon Aviary, completed in 1965 and named after its main designer, the Earl of Snowdon. It rises 80 feet (24 metres) above the water over a steeply sloping site, and its imaginative skeleton of tension cables and massive tubular supports is covered by an expanse of mesh. As they walk along a cantilevered bridge within the Aviary, 20 feet (6 metres) above landscaped pools and past waterfalls, visitors have a treetop view of nesting colonies of sacred ibis, night herons and cattle egrets.

The new Elephant and Rhino Pavilion, also built in 1965, is one of the most striking buildings in the Zoo. Towering above its inhabitants, its heaviness reflects their shapes, and the filtered light inside produces a dramatic effect compatible with the largest living land mammals which are seen highlighted and in silhouette. The building received the 1965 Award of the Royal Institute of British Architects for the best building to have been erected in London that year.

In 1967 the Charles Clore Pavilion for Small Mammals was opened on the north side of the Canal, and it has proved to be very popular with visitors. The enclosed building relies for effect and function on glass-fronted landscaped enclosures, on artificial lighting and air-conditioning, and on the use of plants. In the basement 'Moonlight World' the lighting has been so designed that during the day nocturnal animals become active; at night the light is increased, and the animals rest or sleep. The majority of the animals in the Pavilion have bred, notably lemurs, marmosets, fruit bats, fennec foxes, and many mouse and rat species.

The lastest major developments have been the Michael Sobell Pavilions for Apes and Monkeys in 1972, and the new Lion Terraces in 1976. Both are relatively simple and to some extent adaptable structures, with the emphasis on easy and effective management of both animals and the public. The Lion Terraces include a waterbirds' aviary, where European birds live in a marshland planted with typical vegetation. As in the design of all recent buildings, the accent has been on providing the animals with an environment which is as near natural as possible and where their normal behaviour patterns can develop and they can breed. Within six years of its opening, nine out of eleven species in the Monkey and Ape Pavilions had bred. Three out of seven species in the Lion Terraces had produced young two years after the building was opened.

The adaptability of the Sobell Pavilions was demonstrated when in 1974 two giant pandas, Ching Ching and Chia Chia, were presented to the people of Britain by the government of China. During the period of their enforced quarantine they were housed in the Pavilions, the intention being to build special quarters to receive them afterwards. Because of financial difficulties this has not proved possible, and the animals' temporary quarters, with modifications, have become their permanent and, as it has turned out, quite satisfactory home.

During the period of redevelopment in Regent's Park, equally impressive changes were made at Whipsnade. A number of new buildings were erected and exhibits reorganized on a zoogeographical basis. The most noteworthy building is that for water mammals. Of elegant and simple design, it was opened in 1972 and houses bottle-nosed dolphins in outdoor and indoor pools. Visitors can watch the dolphins under water through the glass sides of the pools. South of the building runs a steam railway which has proved very popular with visitors since it opened in 1970. The train follows a circular course through animal paddocks.

During the Society's history, many buildings have come and gone, the shape and size of enclosures have changed, and the gardens have varied from the rigidly formal to a more natural appearance. Changes in style have mirrored contemporary fashions in architectural thinking, while others have reflected advances in the understanding and management of animals and in the control and prevention of diseases. In recent years there has also been an increasing recognition of the need for animals to be able to behave naturally. The design of recent buildings also reflects a change in general policy. Instead of the comprehensive collections of the past, the Society now concentrates on those animal families or genera which can form self-sustaining groups. In 1966 the Zoo had on view about 120 individuals belonging to 50 species of monkeys and apes; in 1978 there were only 72 individuals of 16 species. In 1966, 17 species of cats were on exhibition; in 1978 there were 8.

With changes in buildings and in the management of animals, the breeding success rate has improved. In 1977, of the 909 mammals living in Regent's Park, 613 (67 per cent) had been born there, and a further 12 per cent had been born in other collections. Nearly 85 per cent of the mammals living in Whipsnade Park were born there. One of the most outstanding successes at Whipsnade is that of the cheetah. Although cheetahs had been kept in captivity for centuries, it was not until 1956 that the first litter was born in a zoo – Philadelphia. The cubs did not survive, but those from a litter born in Italy in 1966 did and were successfully reared. Then in 1967 the cheetahs at Whipsnade produced their first litter, and since then over fifty cubs have been born, most of them successfully reared by the parents. Some of these parents having been bred in captivity themselves, their offspring were the first true second-generation captive-born cheetahs in the world.

Whipsnade was chosen by the Natal Parks Board as the first conservation centre for white rhinos outside Africa. In 1970 a herd of twenty arrived from Umfolozi in South Africa. They settled without difficulty into their new home, and bred so well that the Society was soon able to send surplus rhinos to collections in other parts of the world. Perhaps the most famous conservation story is that of Père David's deer, extinct since the turn of the century in China, their country of origin. In 1956 and 1973 the Society was able to return four young Père David's deer to China. Other notable breeding successes of rare, or rarely bred species, include Przewalski's horse, pygmy hippo, European bison, polar bears, tigers, musk oxen, red-breasted geese, Manchurian cranes, Cereopsis geese, rosy and Chilean flamingos, black-footed and king penguins.

Many species at Regent's Park and Whipsnade have been bred to third, fourth or more generations, and some are now represented entirely by captive-bred individuals. The most noteworthy are the wolf (last imported in 1953), Thomson's gazelle (last imported in 1954), American bison (last imported 1954, with the exception of a new bull in 1963), llama (last imported 1953), axis deer (last imported before 1959) guanaco (last imported in 1960), and barbary sheep (no new blood since the Mappin Terraces opened in 1913).

As the science of zoology developed, so the emphasis of the scientific activities of the Zoological Society of London changed. The formation of the Society came at a period when questions of morphology, physiology and classification occupied the minds of most eminent zoologists. It was a time when anatomists were concerned to show that a knowledge of the structure of an animal, both external and internal (morphology), was essential for the understanding of how that animal functioned (physiology). Linnaeus's system of classification, which assigned animals into a related sequence of categories (species, genus, family, order, etc.), was based on each category having diagnostic anatomical characteristics. The acceptance of his system underlined the need to understand precisely the internal and external structures of animals, so that resemblances between species could be determined.

The founding of the Society also coincided with a period of imperial expansion, of travel and exploration, and animals, some of them new to science and others which had rarely been seen before, were brought back or sent from all over the world to the new Zoo. Zoologists and anatomists were thus provided with the materials and facilities for their research. At the same time the Society provided a forum where the results of researches could be discussed and published. The names of most of the eminent zoologists of the nineteenth and early twentieth centuries can be seen in the Society's publications, together with their discoveries.

With the emergence of the experimental approach in biological sciences in the twentieth century, the emphasis changed from studying the dead to studying the living. There has been an expansion of interests, and two popular facets of natural history have become prominent: ethology – the study of behaviour; and ecology – the study of living organisms in relation to their environment, both animate and inanimate. At the same time there have been developments in the physical sciences which have given us instrumentation of such sophistication that whole new zoological areas have been opened up. The research activities of the Zoo now encompass a wide number of interests – genetical, biochemical, behavioural, nutritional, parasitological, radiological, etc. In 1964 the Society opened the Wellcome Institute of Comparative Physiology for the study of reproductive physiology, and in 1965 the Nuffield Institute of Comparative Medicine for the investigation of diseases in animals, including man. The Wellcome laboratories were set up at a time when the fundamental knowledge of the physiological mechanisms of reproduction were derived almost entirely from observations made on domestic and the common laboratory animals. The Society's extensive collection of mammals and other vertebrates provides a unique opportunity to extend this knowledge. The laboratories, made possible initially by grants from the Wellcome Trust and the Ford Foundation, can accommodate up to fifteen research workers and their attendant staff. The Nuffield laboratories house some twenty professional research workers, together with attendant staff, and provide modern facilities for studying disease in all its comparative forms. There are four departments in the laboratories: biochemistry, infectious diseases, radiology and haematology. Each department has a permanent head with a number of research assistants, research fellows and students. A considerable background of knowledge, gained over the years, on the nutrition, husbandry and diseases of exotic animals has given the research workers a wide scope for enquiry into comparative medicine.

The two research institutes, the animal hospital and pathology laboratory have recently been integrated under the overall title of the Institute of Zoology, of which the Society's four curators, who are all professionally qualified scientists, are also members. There are in addition about twenty postgraduate research workers supported by grants from many sources. Members of the Society's staff published more than a hundred papers in scientific journals during 1976.

It was the intention of the founders of the Society that education, in its widest sense of instructing the public and advancing zoological knowledge, should be one of the prime aims of the Society. A wide range of educational activities has been organized over the years, but it was not until 1958 that an Education Department was established as a separate entity. The Department first occupied the old Cat House, built in 1904, with the interior transformed to accommodate staff and pupils. In the first year 15,760 children attended courses for various grades of academic achievement. By 1977 this figure had quadrupled. So great was the demand for these facilities that in 1975, with the help of a grant from the Wolfson Foundation, a new Education Centre was opened. Its three lecture theatres, classrooms, seminar room, darkroom, audio-workshop and offices provide educational facilities unrivalled in any other zoo. There are courses and symposia for school and university students, Christmas lectures and, for children under eighteen the XYZ (Exceptional Young Zoologists) Club, with its own publication, *Zoo Magazine*. The Centre also includes the Inner London Education Authority's Teachers' Centre for Life Studies, which is used primarily to train and guide teachers in the use of the Zoo for biological instruction.

Another educational activity has been the establishment of courses for keepers in the general management of animals in zoos. The Society's own Diploma Course was started in 1956, but this later developed into a joint activity with Paddington Technical College. The resulting course has become a pioneer of its kind.

Scientific meetings open to Fellows, Associates and interested scientists, are held on the second Tuesday of the months of February to June and October to December. At these meetings, which began as business meetings in the year the Society was founded, the results of new research are communicated and discussed, and exhibits and films of zoological interest are shown.

Such has been the expansion of zoology in this century that specialization has been inevitable. Scientific meetings have often been devoted to one aspect of research, and on occasion a meeting has developed into a symposium where people working on the same research subject discuss and review their findings. One of the first such meetings was held in October 1919 on 'The position and affinities of *Tarsius*'. In 1959 the Society organized the first of a series of international symposia. These symposia and the published volumes which have resulted have been a great success.

Publications have also been a continuing part of the Society's function. Both the *Proceedings* – now the *Journal of Zoology* – and the *Transactions*, have appeared without interruption since 1830 and 1833 respectively. The *Journal* is now in its 186th volume.

The *Zoological Record*, a cross-referenced bibliography which has proved a valuable tool for zoological research workers, was started in 1864 as a joint venture with the British Museum (Natural History), but in 1886 the Society took sole responsibility for it. The growth of zoological literature, especially in the last two decades, has been very rapid, and each year's compilation requires the scanning of some 6,500 journals and books, and now deals with some 60,000 to 70,000 titles, most requiring more than one entry.

In 1939, the first seven volumes of the *Nomenclature Zoologicus* were published. They list, in a chronological series, the names of every zoological genus and subgenus established since 1748, the date of Linnaeus's *Systema Naturae* (tenth edition). Together the seven volumes list about 290,000 entries.

The last major publication started by the Society is the *International Zoo Yearbook*, published annually since 1960. It covers all aspects of captive wild animals, and represents the most up-to-date results of observations and research conducted in zoos and laboratories around the world. Its reference section contains a directory of nearly a thousand zoos, wildlife parks and aquaria of the world, as well as a yearly census of animals bred in captivity and one of rare animals in captivity.

The Society also publishes an informative *Annual Report* and, since 1967, a biennial *Scientific Report*. It owns one of the world's most important collections of zoological literature, consisting of more than 120,000 volumes. The Library also houses the Society's archives, and an invaluable collection of graphic materials including paintings, drawings, photographs and films.

Since its inception the Society has had periods of stability, energetic redevelopment and, occasionally, apparent inactivity. There have been peaks of popularity, criticism and praise, financial crises, public and private quarrels – in short, the normal ingredients of the history of any major and ancient institution. But, above all, the Society has in the course of its development become, because of the breadth and quality of its achievements, a zoological institution which is a model of its kind. It is still a private society with the status of an educational charity, a national institution without annual subvention from the government or municipality. It relies almost entirely on gate money and donations. It has survived because it fills a scientific and social need and because of the dedication of the people who run it.

WEST BERLIN

Der Zoologische Garten Berlin

Colin Rawlins

Apart from the difficult years which immediately followed the two world wars, West Berlin Zoo has always had on display a larger and more comprehensive collection of exotic animals than any other zoo in the world. At the end of 1977 it was exhibiting more than 5,000 animals belonging to almost 1,350 species of birds, mammals, reptiles and amphibians. The comparable figures for San Diego and London, each of which possesses a country 'satellite', were some 4,900 animals of 1,100 species, and 4,700 of 850 respectively. In addition, the West Berlin Zoo keeps important collections of fish and invertebrates. If these are included in the total, the number of animals in the Zoo is more than 12,000 belonging to 2,500 different species.

As is so often the case with the great European zoos, it all began with a royal collection. As early as the start of the sixteenth century, the Electors of Brandenburg kept wild animals for hunting and for the table in an area near Berlin called the Tiergarten – the animal garden. By the middle of the eighteenth century, the Tiergarten had become part of the royal estates of the Prussian kings, Frederick the Great having transformed a section of it into the formal garden that is still there today. Nearly a hundred years later, in 1841, a portion of the park was handed over by King Friedrich Wilhelm IV for use as a zoological garden. This became the Berlin Zoo. He also handed over to the new zoo an excellent collection of wild animals – bears, monkeys, kangaroos, ostriches and cassowaries – which his father Friedrich Wilhelm III, king from 1797 to 1840, had kept on Peacock Island at Potsdam, the royal country residence. The public had been allowed to visit the collection, provided that they did not feed the animals – a remarkably far-sighted prohibition for those days.

Friedrich Wilhelm III had been advised and encouraged in his zoological activities by Dr Martin Lichtenstein, who was the first holder of the Chair of Zoology at the new Friedrich Wilhelm University in Berlin. Dr Lichtenstein was committed to the idea of collecting, both for exhibition and study, new species of exotic animals, many of which he had seen when serving with the Dutch forces in South Africa during the Napoleonic Wars. Here his interests and aims coincided closely with those of one of the greatest naturalists and explorers of the day, Alexander von Humboldt, and the idea of founding a zoological garden for Berlin was in effect a joint venture. After a visit to the new London Zoo in 1833, Lichtenstein returned more enthusiastic than ever. The chance to proceed came a few years later, when Friedrich Wilhelm III died. Friedrich Wilhelm IV, his successor, was far more interested in art than in animals, and as an economy measure he wanted to give up the Peacock Island collection. He therefore readily fell in with the proposal put to him by Lichtenstein and von Humboldt that a zoological garden should be established in the Tiergarten, and that some of his animals should be sent to stock it. In September 1841, an order was therefore made to transfer to a private company, set up to run a zoological garden, the royal pheasantry site of about 60 acres (24 hectares) in the Tiergarten and part of the Peacock Island collection. The King also agreed that the company should be given a loan of 1,500 talers, interest-free for five years, to help cover the initial building costs. The King himself chose forty-seven animals as his first donation and ordered that another two hundred and fifty should be transferred as and when needed at the Zoological Garden.

Thanks to von Humboldt's influence, a further loan of 10,000 talers to help start the Zoo was made by the Prussian government, and more animals were bought. Some of these died, and it was with a collection rather smaller than Lichtenstein had hoped – fewer than 100 species – that the Zoo opened in 1844. The design and layout of the new Garden was chiefly the work of the well-known landscape architect Peter Lenne, who had been responsible for the new gardens of the royal Potsdam residence. For many years he was to be Lichtenstein's partner in the development and management of the new zoo. There were no celebrations at the opening of the Zoo because the event had been overshadowed by an attempt to assassinate the King a few days earlier.

The next few years were difficult ones for Prussia and for Berlin because of the political troubles of 1848 and the prevailing economic depression. The Zoo had the support of the King, but not of the people, nor was its administration by a statutory commission the best that could have been devised. The King's three nominees on the Commission were the most influential, among them Lichtenstein and Lenne, and the former was, to all intents and purposes, Director of the Zoo. The net effect was that at first the Zoo progressed very slowly.

There was little improvement after Dr Lichtenstein's death in 1857, or after the appointment to the Zoo Commission of Dr Hartwig Peters, his successor in the Chair of Zoology at the University. Peters was one of the great nineteenth-century naturalist-explorers, and his work on the animals of East Africa was world-renowned. (One of his early acts was to honour his predecessor's scientific reputation by naming a newly-described African antelope 'Lichtenstein's Hartebeest'.) But he was not cut out for the executive job of managing a zoo. Despite Prussia's swift expansion during the next few years under Bismark's leadership, and Berlin's growth as the capital of the new Germany, the Zoo did not prosper, and its slow development attracted much criticism from the general public and from other quarters. Peters believed that the failure of Berlin's zoo to keep up with other national and civic collections was caused by its top-heavy administration, and he therefore persuaded the governing council to appoint a full-time professional director.

In 1869, Dr Heinrich Bodinus was appointed to the post. He had successfully set up a new zoo at Cologne ten years earlier, and during his term of office in Berlin he succeeded in increasing the numbers of animals from fewer than a thousand in some two hundred species to two thousand in five hundred species. A more lasting achievement was his success in providing new buildings and facilities. The Zoo's financial position had begun to improve in the mid-1860s, and had been greatly strengthened by the raising of new capital funds through shares and loans. It thus became possible to put in hand several improvements: a better water-supply and drainage system, a new restaurant, and three new animal houses – a Large Carnivore House in 1871, an Antelope House in 1872, and an Elephant Pagoda in 1873.

The opening of the restaurant and the inauguration in 1870 of regular concerts by military bands had an immediate effect on the popularity of the Zoo. The new animal buildings, and the new Antelope House in particular – a veritable 'animal palace' as it was then described – also had a great impact on the public, which was certainly increased when the 'animal palace' was used for a luncheon during the 'Meeting of the Three Emperors' – Austrian, German and Russian – on 8 September 1872. A later diplomatic occasion for the Zoo was a grand reception given on 22 June 1878 for the participants at the Congress of Berlin on the Balkan question.

In spite of what today would be called good public relations, the deterioration in the economic situation of Germany towards the end of the 1870s had an inevitable effect on the number of visitors to the Zoo. There were also some unfortunate deaths among the animals, particularly those that were prominently publicized, like that of the popular elephant 'Boy'. The Zoo was also suffering from competition with a quite independent company that operated the Berlin Aquarium. This had been opened in 1869 under the direction of the distinguished naturalist Brehm, who had also founded the Hamburg Zoo and Aquarium. The Aquarium housed birds and mammals as well as fish, reptiles and amphibians. Apes were among its exhibits, and when the Aquarium acquired the first gorilla ever to be seen in Germany, it attracted thousands of visitors who might otherwise have gone to the Zoo.

Among the features of the period were live anthropological exhibitions – Exhibition of Peoples, as they were called. The famous dealer Hagenbeck, of Hamburg Zoo fame, had much to do with this vogue, bringing caravans of African tribesmen with their wild animals on tours throughout Europe. On 6 October 1878, 62,000 people visited the Berlin Zoo to see a group of Nubians from the Sudan, together with camels, elephants, giraffes and rhinos. Other national groups which were paraded through the Zoo were Laplanders, Eskimos, islanders of Tierra del Fuego, Kalmucks and Buddhist priests.

Despite these attractions, the financial affairs of the Zoo had improved very little by the time Dr Bodinus died in 1884. He had increased the collection substantially, and had made the Berlin Zoo world-famous for its fine new buildings. He was popular with Berliners, but because of the financial situation the governing Council decided to appoint as his successor someone who, it was hoped, would be a better administrator. They chose Dr Max Schmidt, who for over twenty-five years had been the Director of the new and very successful Frankfurt Zoo. He died only three years after being appointed but, in the short time that he served, he reorganized the administration and improved the Zoo's finances by enforcing strict economies. He also re-established one of the scientific aims of the Berlin Zoo Society by publishing articles on zoo animals and their management in *Der Zoologische Garten*, a professional journal which he himself had launched soon after the opening of the Frankfurt Zoo.

Dr Schmidt's successor was Dr Ludwig Heck, a young man whose name was to become one of the best-known in the world of zoos. Although only twenty-eight years old when appointed, he had already become an experienced zoo professional through service in the Leipzig and Cologne Zoos – in the latter as Director. He directed the Berlin Zoo for over forty years, and his name was to be carried on by his two sons; one as his successor in 1931 at the Berlin Zoo, and the other, related by marriage to the Hagenbecks, as the first Director of the new Munich Zoo.

The Heck period, particularly that part of it which ended with the First World War, was a brilliant one. The collection trebled in size, in 1913 reaching a quantity and variety never before and never again seen in any zoo. There was more building in the grand style, including a Bird House, Ostrich House and Stork House. The restaurant block was greatly extended and two ornate public entrances were built, one with elephant and the other with lion portals. The Ostrich House, built in the style of an Ancient Egyptian temple, was an example of the zoo architecture popular in the period. The Antelope House was modelled on the design of a mosque and the Elephant House on a Burmese pagoda. At the Fifth International Congress of Zoologists held in Berlin in August 1901, Dr Heck was able to show to the participants the largest collection of wild animals (402 mammal species and 894 bird species) in the world, and all kept 'in the finest buildings'. A professor of the Sorbonne was to describe the Zoo as 'This beautiful and magnificent garden, where the animals live like princes'.

But the reputation of the Berlin Zoo was not then, and had not ever been, based simply on a vast conglomeration of animals in monumental houses. Its management was good and there were some noteworthy breeding achievements, including Père David's deer, wild asses, zebras, yaks, Cape buffalo, Cape hartebeest, bontebok, chimpanzees, and many rare pheasants and waterfowl, as well as emus, rheas, and Demoiselle cranes. In 1908 the birth occurred of an Australian echidna, the first in any zoo. Indian elephants were born in 1906 and 1928, and an orang-utan in 1928, the first to be reared in a zoo.

Some very interesting animals had lived in the Zoo before Heck's time. A Tasmanian wolf (or thylacine), a very rare creature, had been acquired in 1864, and one of the only two live quagga, then extinct in its native South Africa, died in Berlin in 1875. In 1870 Hagenbeck had imported a large consignment of animals from Africa, many of which were bought by the Berlin Zoo. They included an aardvark, and the second black rhinoceros to be kept in any German zoo. There was a unique collection of Asian rhinos: an Indian one had been acquired in 1871, a specimen from Java in 1874, and one from Sumatra in 1879. Royal gifts were frequent. The Tsar of Russia presented a pair of European bison (or wisent) after his visit to Berlin in 1872. In 1881, the Prince of Wales presented two young Indian elephant bulls which he had brought back to London in 1876 after his world tour. They were given to the Berlin Zoo to mark the marriage of his nephew, the German Crown Prince. One of these animals, called Rostum, had killed a keeper in the London Zoo. He killed another keeper in Berlin in 1883.

It was during Dr Ludwig Heck's directorship that an Aquarium was established within the Zoo's boundaries. As already mentioned, the original Berlin Aquarium, which was an entirely separate and independent institution that had been opened in 1869, was a serious competitor to the Zoo, since it kept animals other than fish. The Aquarium was a large building with an elaborate system of artificial caves for exhibiting the animals, and in its time it was the best aquarium in the world. But its site was unsatisfactory and new developments in the centre of Berlin at the start of the twentieth century led to the decision to close it down. A proposal by the Berlin Zoo to build a new aquarium was immediately accepted. Three years later a fine, tall building, the most modern of its kind, with a Terrarium and Insectarium on the upper floors and an Aquarium on the lower, was opened under the direction

Blijdorp Zoo. These new buildings on the edge of Rotterdam were opened in 1940.

ABOVE Many unusual zoo buildings were destroyed in the Second World War. This Elephant House at West Berlin was built to look like a Siamese pagoda. Opened in 1873, it was bombed in 1943 and pulled down.

OPPOSITE La Grande Rotonde, Paris Zoo, built in the form of the Cross of the Legion of Honour, was completed in 1812. It now houses mammals.

ABOVE Goma the gorilla and her baby Tamtam at Basle Zoo. Goma herself was born at Basle, the first gorilla birth in any European zoo.

OPPOSITE ABOVE Baby Polar bear, Paddiwack, now at Alexandria Zoo, was born at London Zoo in 1972.

OPPOSITE BELOW Indian rhinoceros and her eighteen-month-old calf Miris at West Berlin Zoo.

PREVIOUS PAGES An eighteenth-century engraving of the Menagerie at Versailles.

TOP OF PAGE An engraving of London Zoo, from a watercolour by Scharf, *c*. 1836.

ABOVE The new Lion Terraces at London Zoo.

of Dr Oskar Heinroth. It lasted almost thirty years. A direct hit from a bomb on the evening of 23 November 1943 destroyed it completely.

The First World War, and the financial and economic chaos which followed, led inevitably to a run-down in the maintenance of the Zoo and a decline in the number of animals kept. In the winter of 1922–3 the Zoo had to be closed to the public as an economy measure, but the currency reform at the end of 1923 saved the day, and the Zoo survived. At that time the number of species of birds and mammals had fallen to about 500. Helped by his son, Lutz, who had joined the staff in 1923, and who had led two important animal-collecting expeditions to Africa in the late 1920s, Dr Heck soon restored the Berlin Zoo to its position as the leading living museum. By the time of his retirement in 1931, the number of bird and mammal species was back to a figure of nearly 1,300, once again the highest in the world.

Ludwig Heck was succeeded as Director by Lutz. Little building took place during the next seven years, and then war again broke out. The new Director did much to improve the backgrounds of dens and enclosures with some excellent artificial rock-work. New enclosures for bears and wolves, as well as the monkey 'rocks', were constructed during this period.

If the First World War had a disastrous effect on the Zoo the Second all but put an end to it. Animal losses were catastrophic. Only 90 of about 3,600 birds and mammals survived the RAF and American bombing, the shelling during the final attack on Berlin and subsequent looting. A Greek tortoise was the sole survivor of the extensive collection of animals which had filled the great Aquarium building before the war.

The first casualty from bombing occurred on the night of 8 September 1941, when an eland cow was killed. The main losses, however, came in the big air attacks at the end of 1943 and the beginning of 1944 when, in four night-raids, the stock of the Aquarium was almost completely wiped out, and several of the biggest and most important animals killed, including seven elephants, a rhinoceros, a chimpanzee, and an orang-utan. It had been decided earlier that some of the animals should be evacuated to other zoos, but progress with this plan had been slow. It was pressed forward after the big raids, and in the end some five hundred animals were sent away. Only one of them, a female giraffe which had been sent to Vienna, was returned to Berlin after the war. In the Zoo itself, only ninety-one animals belonging to forty-five species were still living in corners of the elegant Tiergarten over which battle had raged. One of the survivors, a hippopotamus, was to father thirty-nine babies.

The increase in the size of the collection is the most remarkable of all the Zoo's post-war achievements. By 1956, when the first post-war

Director, Dr Katharina Heinroth, widow of Dr Oskar Heinroth, retired, the number of birds and mammals had risen to 2,000, representing more than 450 species. Under the energetic leadership of her successor, Dr Heinz-Georg Klös, these figures almost doubled in the next ten years, mainly by additions to the bird collection. Today, the West Berlin Zoo once again holds more animals than any other zoo.

But the period of post-war reconstruction was marked by a very dramatic start. Three years after the war ended in 1945, just when some sort of order was being restored to the Zoo grounds, the gigantic 'flak' tower in the Tiergarten next to the Zoo was demolished by the occupying powers. The explosion caused almost as much damage in the Zoo as had the war. Despite this devastating new setback, the work of reconstruction was begun again, mainly because of the tenacity of Dr Katharina Heinroth and her able lieutenant, Werner Schröder, who later became the Director of the rebuilt Aquarium.

Dr Heinroth had managed to rebuild the Antelope House, the 'animal palace' in which monarchs had been entertained, and to build new houses for elephants and hippos. But the main developments of the post-war period have been due to Dr (now Professor) Klös who, in 1957 at the early age of twenty-nine, started to build animal houses and enclosures in accordance with the most up-to-date knowledge and experience, to develop a comprehensive and healthy collection of animals, and to re-establish the international prestige of the Berlin Zoo.

Money and land were essential prerequisites, and loans and grants, including the proceeds of part of the sales of postage stamps commemorating the Zoo's famous Director, Dr Ludwig Heck, have helped with the collections. The city has provided the capital for all the redevelopment that has been carried out continuously since the early 1950s. The company owning the Zoo (for it still operates in the way it did when set up in 1844) has covered running costs by charging realistic prices for entry.

An adjustment of the original boundaries of the Zoo was negotiated with the city authorities in the mid-1950s. Land to the south was surrendered to urban development in exchange for some to the north, including the area which had contained the demolished flak-tower. These adjustments left the Zoo with an area of nearly 90 acres (36 hectares), large by the standards of most city zoos. Part of the new land has been used to construct one of the best service areas of any zoo in the world, with excellent workshop and storage facilities.

New buildings went up in quick succession. The big Aquarium building, of which only the foundations remained intact after the war, was completely rebuilt between 1952 and 1960. A magnificent Bird House, one of the largest exhibits of its kind in the world–with an area of three-quarters of an acre (0.30 hectares) and containing 181 aviaries–was opened in 1962. A house for rhinos and tapirs followed in 1963; big new open enclosures for bears in 1969; and over the period 1970–5 the accommodation for lions, tigers and other big cats was completely modernized with an underground exhibit for nocturnal animals being incorporated on the same site. During this period many other enclosures and new service buildings were built, including offices and public amenities such as restaurants. In 1977 the Tropical House, designed for forest-dwelling animals, went up.

There are few gaps in the West Berlin collection of animals. Antelopes and deer are unusually well represented, and the Zoo contains what is probably the best collection of wild cattle anywhere in the world, including such rarities as anoas, the small cattle from the island of Celebes, and the gaur and banteng, the larger wild cattle from south-east Asia. There are large collections of birds of prey, including owls, and waterfowl and cranes. The public's favourites include a pair of southern elephant seals from the islands of the Southern Ocean, gorillas and orang-utans.

A large percentage of the creatures in the Berlin Zoo were born there, including zebras, hippos, both species of the small muntjac (or barking deer) from Asia, banteng and gaur, orang-utans, and seals; and among the birds, flamingos of two species, snowy owls and Hawaiian geese. Great bustards and mountain anoas were bred for the first time in any zoo, and several other species for the first time in Germany or Europe. The Zoo has also played its part in the international effort to save such animals as the European bison and the Père David's deer, and helps in conservation by keeping the studbook for the black and white rhinoceros and the gaur.

The population of West Berlin is only about two million, and the fact that the Zoo has nearly two and a half million visitors every year is not only a tribute to its quality, but an indication of the affection with which it is regarded by the Berliners, and of its role as a community centre. The vast crowds that gather in the Berlin Zoo over Whitsuntide, some from eight o'clock in the morning, to listen to the traditional concert and to wander round this great show of animals, must have an enormous sense of pride for what they, the West Berliners, have managed to build on the foundations of a ruined zoo.

ANTWERP

Zoo Antwerpen

Michael Brambell

Before 1830, when it became an independent state, the Belgium of today had been successively ruled and fought over by Spain, Austria, France and Holland. It was certainly worth fighting for. It was rich in natural resources, and trade had made it even richer. Equally, contact with other peoples had made it a focal point in the development of European culture. Both Brussels and Antwerp were powerful centres of commerce, and the latter had been a major port for generations. Ships from distant parts of the world carried to it not only articles of trade, but also creatures and plants, many of which were recorded for posterity by the brush and pen of great artists. Adrian Collaert, at the end of the sixteenth century, made magnificent engravings of strange as well as indigenous birds. Exotic animals, which must have arrived by sea, figure in many of Rubens's landscapes, of which *Earthly Paradise*, now in the Hague, and painted jointly with Jan Breughel – like so many of his other canvasses in which animals appear – depicts a rich array of animals from Africa, the Far East and South America.

In 1841, eleven years after the new state had been proclaimed, some prominent citizens decided to establish a permanent exhibition of exotic animals and plants that were being brought into the country. Amsterdam had just started its zoo, and so had other European cities. In 1843 Antwerp therefore followed suit with the foundation of the Société de Zoologie which King Leopold I, Belgium's first monarch, allowed to call itself 'Royale'. Even before this the founders had bought the first 4 acres (1.6 hectares) of the present zoo site, and also a collection of animals, from one Jaak Kets, who was then installed as first Director of the new menagerie. Ships' captains were asked to bring back any interesting animals that they came across on their voyages, and one, a Captain Meyer, 'donated' to the Society a young African, who was 'taken on' to look after the birds. This man, called Jefke, became a well-known Antwerp character, marrying a local girl and eventually becoming one of the Zoo's gatekeepers. Among the first animals which the Zoo acquired were monkeys, flamingos, a bear, peafowl, pheasants and an ostrich. To reach the Zoo, visitors had to walk through a postern gate in the ancient ramparts of the city, cross the old moat by a wooden bridge, and then step over the rails of the main railway line to Holland.

Whenever the opportunity arose, the Society bought up small parcels of land to add to its first purchase, and by 1888 it had acquired most of the 25 acres (10 hectares) which the Zoo now covers. The first major building went up in 1854. Built in the style of an Egyptian temple, and with a façade incised with hieroglyphs, some regard it as the most striking nineteenth-century building still standing in any zoo. It was put up to house elephants, rhinoceros, giraffes, camels and zebras. A

Carnivore House, with 16½-foot (5-metre)-high heraldic lions at its entrance, was built in 1879. The lions are still preserved in the Zoo, but the House has long since gone. A house for hippopotamuses followed in 1881. Then, in 1897, came a vast concert hall, with seating for two thousand, and in 1901 a grand new gateway to the whole Zoo, crowned by a bronze statue of a boy riding a camel. Today the statue stands on top of the Society's offices, close to the main gateway. The only further building of note that went up before the start of the First World War in 1914 was a combined Aquarium and Reptile House, which is also still in use.

Up to this time, Antwerp's population was little more than a quarter of a million, and while commercially more important than Brussels, it was the latter, the country's capital, which was the seat of the Belgian Royal Academy of Arts and Sciences, an institution that had been founded in the latter part of the eighteenth century. Antwerp was still what today we would call a provincial city, lacking even a university. It had, however, started a civic zoological society, and was also the home of Belgium's Royal Academy of Fine Arts. It was not at all surprising, therefore, that the latter joined forces with the Royal Zoological Society to transform the grounds of the new Zoo into the city's cultural centre. And that is what it remains right up to the present. The concert hall was used not only for meetings of the Zoological Society, but also for musical performances and exhibitions of paintings. Until the 1920s, these added at least as much distinction to the Zoo as did its animal collections.

It was only after the First World War that Antwerp became prominent in the world of zoos. In 1901 Sir Harry Johnston, a member of the Council of the Zoological Society of London and a Foreign Office official who had served several tours of duty in Africa, discovered a new animal in the Belgian Congo. It was the okapi, a hoofed animal midway between a giraffe and an antelope, which had never before been seen by any European. The circumstances of the discovery were somewhat gruesome. News had reached London that some Ituri pygmies were going to be forcibly abducted and taken to Paris to be put on show at an international exhibition. Sir Harry was instructed to intercept the European who was committing this outrage and to return the people to their forest home. This he succeeded in doing, but on the return journey from the coast one of the pygmies died. Knowing that skeletons of pygmies were rare, he decided to bury the body in an ant-heap until the bones had been picked clean. This took no more than a few days, and the remains were then boxed and sent to the British Museum in London. The journey to the forests was resumed, and it was there that

Sir Harry obtained the skin of an okapi, which he also sent to London. His description of the animal, new to science, reached the zoological world before the skin, causing a considerable stir, and needless to say also stimulating an immediate search for living specimens.

The okapi lives only in the forests of the Congo basin within the boundaries of what is now Zaire. The first specimen that survived for more than a few days after being captured was sent to the Antwerp Zoo in 1919. Neither it nor a few others that were exported to other zoos at about the same time lived more than a few weeks. Then in 1928 a batch of five reached Antwerp, and a few survived. The creature remains a rarity, and in 1975 there were only fifty-one specimens in sixteen collections throughout the world, of which forty had been bred in captivity.

Then a second animal previously unknown was discovered in the Belgian Congo. This time it was the Congo peafowl, a fascinating bird which also lives only in the forests of central Africa, and which is separated by thousands of miles from eastern Asia where all other peafowl are found. Again the Antwerp Zoo had first claim on captive specimens, which have bred there most successfully. Before the ancient adage – *Ex Africa semper aliquid novi* (Always something new from Africa) – could be proved right once again, the Second World War had broken out.

During the five years that the war lasted the Antwerp Zoo all but came to an end. Soon after the start of the German occupation in 1940 it became clear that the carnivores would have to be killed because they could not be provided with food. Then it was the turn of those animals that could be killed to provide meat for human consumption. The rare creatures which were preserved lived in a state of near starvation until the liberation of Belgium in September 1944. But that was not the end. Antwerp then became a constant target for flying bombs, which the Germans aimed at the port in an attempt to obstruct its use by the advancing Allied armies. The Zoo was hit several times. The Winter Garden was destroyed, as was the south-west corner of the Zoo. When hostilities ended in May 1945, the Zoo was minus many buildings, without stock, without money, and without several of its experienced staff, including its Director, Michael L'Hoest, who died as the war ended. He was the third of his family to serve in the post, his father having succeeded his grandfather who had been appointed in 1888.

Reconstruction began almost immediately. The Winter Garden was rebuilt for birds, and a new Great Ape House erected. Work started in 1968 on a new complex called the Jubilee Project, which not only replaced the old Carnivore House, but also introduced to the Zoo a

Dolphinarium, an Insect House and a Nocturnal House. The Reptile House has been completely redesigned, and a new Monkey House finished.

The entrance gate is in the north-west corner of the Zoo site, adjacent to the main office building. This also houses the concert hall, entered directly from the street, a museum of natural history and a planetarium. Most of the large buildings in the Zoo are round the edge of its triangular site. On the eastern side of the triangle are small houses and enclosures, including the one in which okapi are kept. There is also a children's zoo, some water mammal exhibits, a paddock for white rhinoceroses, and a house for Père David's deer and South American llamas and alpacas. Along the south side are paddocks for the larger hoofed animals, as well as an aviary for wading birds. In this part of the Zoo are the Aquarium and Reptile House, with their imposing classical porticoes. There are several tidal tanks in the aquarium section, in which the water rises and falls during the day, enabling inter-tidal fauna to be exhibited. Since well over half of all the major zoological groups are marine and since more than half of these forms live on or near the sea-shore, this simple expedient has allowed a whole new range of animal life to be exhibited. By another simple device, that of dividing a tank into two by means of a glass panel, visitors get the impression that predator and prey fish are kept together. Corresponding techniques have been used in other zoos with mammals – in New York, for example, the use of concealed moats has allowed lions and antelopes to be shown in what seems to be one paddock.

The western side of the Zoo is dominated by the Jubilee building complex, which can be approached from the Reptile House over a bridge that spans an enclosure holding mountain sheep and cattle. The path from the bridge leads to the roof of the building and a garden, on either side of which are housed birds of prey. Beyond the roof-garden is a Nocturnal House, not as big as the one in the London Zoo, but incorporating and improving upon many of the ideas in London's 'Moonlight World'. Close to the Nocturnal House is a small collection of insects, and then the Dolphinarium, which follows fairly standard lines. A vast amount of behind-the-scenes activity is needed to maintain a dolphinarium. The filters have to handle a huge flow of water, waste material has to be removed, the salinity maintained, and acidity controlled.

The central part of the Zoo site is given over to a few major exhibits and several minor ones. Near an attractive wildfowl lake and the 'Egyptian temple' is the Hippopotamus House. There is also a new house for exotic birds, where the Congo peafowl can be seen, although the bulk of the breeding stock is not on view.

In 1936 the Antwerp Society bought a 90-acre (36-hectare) estate called Plankendael, 20 miles (32 km) to the south-west of the city, for use as a country zoo. It is lavishly planted and well laid-out, with several large paddocks and aviaries. There is also an ornamental lake. Many breeding groups of birds and mammals are kept at Plankendael, including pudu (a rare South American antelope), bongo, Przewalski's horse, Indian rhinoceros, European bison, Dall's sheep, and monkey-eating eagles. But the site is subject to flooding, the reverse of the situation at De Zegge, a nature reserve which the Society bought in 1951. De Zegge is in the centre of an extensive wetland which has been drying out in recent years, mainly because of drought, but partly also because of drainage and urban development schemes. The Zoo is trying to alert Belgium to the fact that a unique piece of country is being transformed.

In its immediate veterinary work the Zoological Society cooperates with the Institut Royal des Sciences Naturelles de Belgique and also with the Universities of Antwerp and Liège, and with other appropriate Belgian organizations. The Society is also associated in its wider interests with the Universities of Antwerp, Brussels, Liège and Namur, with the Belgian Institute for Tropical Medicine, the Museum of Central Africa, Tervuren and with many other laboratories at home and abroad. The Society publishes occasional monographs on selected subjects, as well as its own scientific journal, *Acta Zoologica et Pathologica Antwerpiensia*. It also issues a popular journal, *Revue Zoo*, which contains useful articles about the Zoo and about general biological subjects.

The teaching staff of the Zoo caters each year for about 80,000 students from schools and colleges in Belgium and the Netherlands. The other educational service rendered by the Society is by way of its Museum of Natural History and the Dr Van Heurek Museum of Zoological and Botanical Specimens. The Society possesses a particularly comprehensive collection of diatoms, and a collection of early microscopes, including one that belonged to Anthony van Leuwenhoek (1632–1723), the brilliant Dutch microscopist. The Planetarium is also one of the Society's cultural ventures, as are the symphony concerts which have taken place regularly since they were started by Michael L'Hoest when he was Director more than seventy years ago. The Society also owns and manages two art galleries, the Alfred Ost and the Charles Verlat, both of which are in almost constant use.

Although in receipt of municipal, provincial and national grants and subsidies, the Society remains an independent body. It maintains close links with other well-managed zoos in the exchange of breeding stock and new information. Its okapis are managed as a single breeding stock with those of Rotterdam Zoo; its pudus are managed with those of

London, Cologne and Rotterdam; its pygmy chimpanzees with those of Stuttgart; and its Congo peafowl have been sent to several other zoos to set up daughter groups.

The Director, Walter Van den bergh, retired in 1978, having shouldered the main burden of the post-war resurrection of the Zoo, which he passed on to his successor in a more thriving state than it had ever been.

ROTTERDAM

Diergaarde Blijdorp

With a note on Amsterdam

Colin Rawlins

There are two well-known zoos in The Netherlands, one in Amsterdam, the other in Rotterdam. The Amsterdam Zoo was founded in 1838, nineteen years before its sister institution, and while smaller than the latter, had a much more distinguished beginning. It was launched by the Royal Zoological Society of Holland, formally known as Natura Artis Magistra, and the zoo was soon given the popular name of Artis. Associated with it are a library, and an ethnographical and natural history museum. The Zoological Institute of the State University was until recently close to the Zoo, which in 1928 built its own laboratory for research into animal behaviour. For some years this was directed by a distinguished scholar, Bierens de Haan. The scientific and educational credentials of the Amsterdam Zoo could hardly have been more respectable.

The Rotterdam Zoo did not enjoy so prestigious a beginning. Unlike most European civic zoos which were started in the nineteenth century, its origin owed nothing at all to the efforts of influential men who were concerned to encourage the spread of knowledge, nor was it favoured by any royal patronage. It all began with two railway employees who kept exotic animals as a hobby and who, in 1855, were allowed the use of a piece of land alongside a railway track that ran from the town. Within a year, they had been joined by some five hundred other animal enthusiasts in an association which the Rotterdam City Council then formally registered as a zoological society. Early in 1857 Henri Martin, a well-known animal trainer, was appointed the first director of a zoological gardens which the Society proposed to establish. Each member had to subscribe 10 guilders a year – a relatively large sum in those days – and the Society also raised a loan to help cover the cost of developing its land. The landscape architects whom they charged to produce a plan proposed that the 35-acre (14-hectare) site should be laid out as a combined botanical and zoological garden. This combination has remained a feature of the Rotterdam Zoo up to the present day.

As a private society, the privileges of the Gardens were reserved mainly for members, the general public being admitted to the Zoo on payment only a few days in each year. In spite of the financial penalty this imposed on the Society, it was not until after the First World War that the number of days on which non-members were allowed to enter was increased. Nonetheless, in spite of financial difficulties, the Society managed to develop its property until by the turn of the century it was becoming one of the major zoos in the world. The Society also succeeded in putting up at least one striking building – a large conservatory more than 100 feet (30 metres) high with a domed roof topped by a figure of the goddess Flora.

Rotterdam itself was developing much faster than was its Zoo. By the end of the First World War, the latter had consequently become enclosed by new buildings in the city's centre. The site by now belonged to the Society which, in November 1929, was informed by the City Council that no further buildings would be permitted in the Zoo. It was too close to what was to become the main railway station. The Society decided to build a new Zoo.

A 'green-field' site, 40 acres (16 hectares) in extent, was found just beyond what was then the edge of the city. The opportunity of designing a completely new zoological and botanical garden on virgin land was leapt at by the firm of architects which was selected for the task. With the proceeds of the sale of the valuable old site, and with the help of a loan from the city, building began in 1938 on the polder called Blijdorp (Happy Village).

But before the site of the old zoo – which was still open to the Society's members and to the public – had been properly cleared, the Second World War started. The process of demolition was then abruptly completed by the Germans on 10 May 1940, when Rotterdam was the target of a fierce air attack. Fortunately the new zoo at Blijdorp was not touched, and two months later part of it was opened to the public. The official opening followed some six months later, when it was decided that the public, like the members of the Society, should be allowed, free of charge, into the still-unfinished Gardens. The result was that, for the rest of the war, during which the new zoo suffered practically no damage, it became a gathering place for the citizens of occupied Rotterdam. Its great Riviera Hall served as a combined theatre and community hall for meetings, concerts and even auctions throughout the war years.

The Zoo's wartime activities and attractions were to pay a handsome dividend. In 1948 the City Council agreed to provide an annual subvention towards the maintenance of the Zoo. In 1971 another agreement established a formula which would provide a subsidy to cover the Zoo's financial deficit. Today some 70 per cent of the Zoo's outlays, both for operations and capital development, is covered by the city's subsidy. In 1957, on the hundredth anniversary of its foundation, the Society's achievements were recognized by the grant of the appellation 'Royal'. Its title today is Stichting Koninklijke Rotterdamse Diergaarde – the Royal Rotterdam Zoological Garden Foundation.

The main buildings of the Zoo lie in the central, elliptical part of the Gardens, with paddocks and service facilities on the periphery. The site is flat and sandy, and while originally treeless, it now has a beautiful wooden appearance as a result of extensive planting since the war. The elegant Riviera Hall, which is the Zoo's centrepiece, has been altered and modernized since 1970. The areas between the main building and its two wings have been roofed over, so that the building is now not only a winter garden, but also a winter zoo where, whatever the weather outside, visitors may spend an interesting and comfortable day under cover. The central part of the main hall is a luxurious garden of tropical plants. Around it Australian mammals, African birds and South American fish are attractively exhibited. To the left of the main hall are quarters for elephants, rhinos and hippos. There is also a circular house for gorillas, chimpanzees and orang-utans. To the right of the hall there is the tropical bird house, and the round conservatory full of tropical plants, including orchids and the spectacular *Victoria regia* water-lily, the largest species of water-lily in the world.

To the south of the Riviera Hall is a large house for lions and tigers, and to the north of the Hall a large lake with the circular exhibition hall or Diorama which stands on the site of an old sea-lion pool. At the northern boundary of the Zoo are the Giraffe House and Okapi House and their enclosures.

The collection of animals in the Rotterdam Zoo, numbering at the end of 1977 just over 2,300 birds, mammals, amphibians and reptiles of some 600 species, as well as about 600 fish, continues to reflect the Netherlands' interest in, and association with its former empire in south Asia – the natural habitat of the orang-utan, the tiger, the banteng wild cattle of Java, and birds-of-paradise. More than 500 tigers of three sub-species – Bengal, Sumatran and Siberian – have been born in the Rotterdam Zoo, an achievement unequalled elsewhere. Tigers, like other large cats, are expensive to keep, but Rotterdam continues to breed them in order to ensure there is a constant supply available for other zoos.

The Zoo's collection of hoofed animals also includes the rare okapi from central Africa, the gemsbok from southern Africa, and the pudu, a species of very small deer from South America. There are seven okapi, all but one of them born in captivity. Only Paris and Bristol have larger herds. The banteng is also now rare in the wild, but it too is breeding well in Rotterdam. Other notable species of hoofed animals that breed regularly in Rotterdam are Przewalski's horse, the onager, Père David's deer, zebra and Malayan tapir.

Rotterdam had the first group of orang-utans to breed regularly in a zoo. Beginning in 1949 when the first birth took place, there have already been twenty-four births. At present there are ten animals in the collection, two having been born in 1977.

What is probably the finest collection of birds of paradise to be seen outside their natural habitat of New Guinea and northern Australia is kept in the tropical-house section of the Riveria Hall. Many other tropical rarities have been kept in the Zoo, including the Congo peafowl which is found only in the deepest tropical forest of eastern Zaire and which was discovered only in 1936. Virtually nothing is yet known of its life story, and much interest was aroused in 1961 when some were bred in Rotterdam.

During the winter of 1977–8 an exhibition – the Notogaea Exhibition – was arranged in the Riviera Hall and its associated buildings to display a collection of animals and plants from the Australasian zoogeographic region of that name. There were twice as many visitors to the Zoo that winter than in any previous year.

The Diorama Building serves as a natural history museum and as the headquarters of the Zoo's Education Department, which is concerned mainly with the instruction of visiting school children and the organization of exhibits in the museum.

Rotterdam Zoo began in a unique way, as a most private of private ventures. It has evolved into one of the finest zoos of modern Europe. That it will assuredly remain, now that it has become what is in essence a public service paid for out of public money.

FRANKFURT

Zoologischer Garten der Stadt Frankfurt-am-Main

With notes on Munich and Stuttgart

Colin Rawlins

There is little that is surprising in the fact that Alexander von Humboldt, one of the foremost German scientists of the late eighteenth and early nineteenth century, played a significant part in encouraging the establishment of German zoos. He was a man of great intellect whose explorations of South America, from which he brought to Europe notable collections of animals and plants, had made people treat him as second only to Columbus in the discovery of the New World. His influence with the Prussian court, of which both he and his father were Chamberlain, had saved the Paris Zoo during the Napoleonic Wars, and had helped enormously in the foundation of the Berlin Zoo in 1844. Other German cities, including Frankfurt, were soon to follow Berlin.

What is extremely surprising is that the foundation of a zoo in Frankfurt owed a great deal to the efforts of that other well-known German of the time, the Kantian philosopher Schopenhauer, who today is remembered more for his pessimistic and ascetic outlook than for the views he embodied in *The World as Will and Idea*. Schopenhauer was an outspoken animal-lover who made clear his feelings about his fellow-men when he wrote, 'Men are the devils on earth and animals the tormented souls.' With other influential citizens, he persuaded the city authorities of Frankfurt to lease 15 acres (6 hectares) of wooded land in the centre of the city as a zoological garden. The site was particularly suitable, including as it did a large building, which could be used for social gatherings, a dwelling house and out-houses. A company was formed to raise 50,000 guilders by the sale of 250-guilder shares – guilders being the currency in use before the unification of Germany in 1871. Most of the leaders of Frankfurt society, including the Rothschilds, contributed. More than 80,000 guilders were collected before the zoo was opened. A three-man team was sent to study the management of animals in Antwerp and in other Belgian zoos, and at the same time an animal collection was assembled from a variety of sources, some creatures being bought while others were presented as gifts. The list of donors included some noble names, including that of the King of Württemberg.

Opened to the public on 8 August 1858, the Frankfurt Zoo was an instant success. Unlike many of the other great zoos that were started during this period and which grew out of small private collections, Frankfurt was the fruit of careful planning. It started with an excellent site, enough money, a good collection of animals, and powerful local support. The Founding Committee of the Zoo had set out to provide a place which would be 'a living textbook of natural history'. Their aim was quickly achieved.

The first Director, Herr Leven, had previously run a zoological museum. He had been responsible for most of the planning, but soon left as he did not get on well with the Committee. In 1859 his place was taken by Dr Max Schmidt, who had joined the staff a few months earlier as the Zoo's veterinary surgeon, and who was to remain Director for twenty-six years.

Soon after his appointment, Dr Schmidt launched the journal *Der Zoologische Garten*, the first to appear in the German language on subjects dealing with the management of zoo animals, and still today the main journal of its kind in German. But his main preoccupation was to find a permanent site for the Zoo. The owners of the land on which the Zoo started were unwilling to sell, and made it clear that the lease could not be extended indefinitely. The area was in a residential part of the city which was becoming more and more fashionable as Frankfurt prospered. Above all, the site was not big enough.

The Pfingstweide (Whit Pasture), just outside the city, was eminently suitable. It had been used for fairs and other public gatherings, including the historic occasion when Napoleon reviewed his troops after the nearby battle of Hanau in 1813. It had also been the meeting-place of mobs in the revolutionary year of 1848. The city administration was agreeable to leasing the site but financial difficulties caused mainly by the Austro-Prussian and Franco-Prussian wars of 1866 and 1870–1 delayed a final agreement until 1872, when a ninety-nine-year lease was granted to a re-established Frankfurt Zoological Society.

Building was started immediately, and by the opening date of 29 March 1874 much had been done. Some of the old buildings, including the Elephant House, were moved piece by piece from the original site, while others, such as the Carnivore House and the Monkey House, were built afresh. Two innovations were an Aquarium and a large 'community' building, which contained a restaurant, theatre and meeting-rooms. This building, with its neo-classical façade, handsome proportions and broad terraces, was ready by the end of 1876, and immediately became the social centre of Frankfurt. Architecturally, it was regarded as one of the finest buildings of its time. Although almost totally destroyed in the Second World War, it has been rebuilt and today stands almost unchanged.

The Aquarium was Dr Schmidt's greatest achievement. There were few other large aquaria in the world at the time, and the only one in Germany was the one which Alfred Brehm had built in Berlin. Frankfurt's Aquarium set a new standard in layout. It had fourteen large fresh-water and salt-water tanks furnished with the biggest and strongest glass fronts that had ever been made. The pumps and filtration apparatus were technically better than those used in any other aquarium. The main water tank for the system was set in an ornate tower which, until its destruction by bombs in 1943, was a feature of the site. The Aquarium was opened in July 1877, and a few months later it was deemed worthy of a visit by the Kaiser.

In 1885 Dr Schmidt, who had moved to the Berlin Zoo, was succeeded by Dr Wunderlich who three years later left to take over at Cologne. He was succeeded by Dr Haacke, a zoologist known for the studies he had made of the reproductive system of the Australian echidna. Five years later Dr Haacke was followed by Professor Adalbert Seitz, a world authority on butterflies and a former ship's doctor, whose travels had provided him with a wide knowledge of the animal world, which he was able to put to excellent use at Frankfurt. During the thirteen years he was Director, he considerably increased the collection, in both variety and number. Professor Seitz also made changes in the Aquarium. Sections for reptile and insect displays were built on the roof of the original building, while the interior was altered to provide for an ornamental fish gallery and a simulated tropical marshland for crocodiles.

Professor Seitz was succeeded by his assistant, Dr Priemel, who held the post for thirty years. It was during his time that the protection of wild animal species threatened with extinction became part of Zoo policy. Dr Priemel had already started a studbook for captive European bison (or wisent), an animal which was becoming scarce in the forests of eastern Europe, its natural habitat, and in 1923 he helped in the formation of the International Wisent Protection Society.

In the second year of the First World War, the Frankfurt City Council had taken over the financial responsibility for the Zoo. But the years of economic depression which followed the peace, the political upheavals of the late twenties and thirties and finally the outbreak of the Second World War, prevented any further developments.

The city of Frankfurt was one of the most heavily bombed targets of the war, and on 18 March 1944 the Zoo's neat lawns, ponds and fine buildings were transformed into a wasteland of craters, rubble, and charred and broken trees. The vast Community Building was an empty shell. Its destruction was due not only to bombs but to the explosion of ammunition that had been stored in it for the anti-aircraft guns mounted on the roof. The Aquarium and most of the other animal buildings were almost completely destroyed. Ironically, the old 'Bear Castle' which dated from 1858, and which had long been overdue for demolition, was spared. The devastation in the Zoo was matched by that in the city as a whole, and the chances of rebuilding the Zoo when the war ended must have seemed remote to the people of Frankfurt. That it was rebuilt was due more than anything else to the determination of Dr Bernhard

Grzimek, a veterinary surgeon who happened to be in Frankfurt when the war ended, and who was immediately appointed chief of police.

His real interest, however, was in the Zoo which the city authorities, with a proper sense of priorities, wanted to close lest resources were diverted from more pressing needs. But with industry in total disarray, and voluntary labour available to help clear the city of the wreckage of war, Dr Grzimek decided to clear up the Zoo. Bomb-craters were filled, the smashed ponds relined, fences patched and cages mended in what remained of it. Fewer than a tenth of the number of animals had survived the bombing and subsequent looting, yet by July 1945, only two months after the war had ended, Dr Grzimek was able to persuade the American Military Government to re-open the Gardens. The response exceeded all expectation. Deprived of almost any other form of recreation, even of swimming in the River Main because of the risk of disease, the people of Frankfurt flocked to the Zoo. In the ensuing six months the Zoo received three times more visitors than in any comparable period in the thirties. To add to its attractions, Dr Grzimek brought in actors and variety artists, ballet dancers and musicians, who performed both in the ruins of the Community Building and in the open on makeshift stages with makeshift sets. Circuses based themselves in the Zoo, and the only cable railway left intact in Germany was moved from the Black Forest to provide an extra form of entertainment.

The Zoo had two million visitors in 1946. More came in the following year, and enough money had been earned to start re-building. Theatrical and other entertainments in the Zoo were stopped and the Frankfurt Zoological Society was once again able to concentrate on its main purpose. Financial stability was assured when the Frankfurt City Council resumed regular grants for the maintenance of the Zoo, and by money raised by a 'Society of Friends of the Zoo'. (This Society was later renamed 'The Zoological Society of 1858, Frankfurt-am-Main'.) By the time of the hundredth anniversary in 1958, most of the damaged buildings had been repaired or replaced. The rebuilding of the Zoo meant that new ideas about animal exhibition could be readily pursued.

The two most significant buildings in the Zoo from the historical point of view – the big Community Building and the Aquarium, now called the Exotarium – were re-opened in 1957. The former now has the only nineteenth-century classical façade that remains in Frankfurt. The building itself was completely modernized and houses the main restaurant facilities, the administrative offices, a theatre (known as the 'Little Theatre in the Zoo'), and spacious meeting rooms. The Exotarium is also a very different and larger building than the original Aquarium, containing penguins, reptiles and insects as well as fish.

Views of various kinds of landscapes, including a polar scene and a tropical river bank, are one of its main features. The Bird House was finished in 1961, the Monkey House in 1962, the Gorilla Enclosure in 1970, and recently a nocturnal animal house (Grzimek Haus), has been completed. The Bird House was one of the first in which birds were free to fly about either uncaged or behind glass and against a simulated natural background.

The outside enclosure for gorillas struck a new note. Armoured glass rather than the usual moat or fence was used to contain the animals, so giving the illusion of an open display while providing protection for the animals from bad weather. Glass had long been used as a barrier in indoor cages – indeed, the Frankfurt Zoo had used it in 1871 to enclose a chimpanzee den – but it had hardly ever been used before for outside enclosures.

The Frankfurt Zoo had started in 1858 with a comparatively large and varied collection of about 1,000 animals belonging to 350 species, including monkeys, camels, llamas, zebras, several deer and antelopes, seals, otters and many birds, including parrots. By the end of the century the collection had grown considerably both in numbers and variety, and at the start of the Second World War, with more than 3,000 animals, it was second only to Berlin among German zoos. Among the few hundred animals which survived the war were two chimpanzees and a hippopotamus called Toni, who fathered sixteen young and died only in 1977. Dr Grzimek was very soon obtaining more animals, among the early additions being some circus animals and two lions, which had been privately owned by an actor and which had spent the last months of the war in a slaughterhouse in Passau. It is an interesting reflection on conditions in Germany at the end of the war that any of these creatures should still be living; there were few such animals in other countries which were caught up in the hostilities. By 1964 the Zoo housed nearly 5,000 animals of 900 species. The corresponding figures today are 3,500 of 650 species, a reflection of the policy being adopted by most of the world's leading zoos of having fewer species on exhibition.

The Zoo has also bred several rare animals: for example, the first okapi, the first black rhinoceros and the first bongo antelope born in Europe. In company with the Columbus Zoo in the United States, and Basle in Switzerland, Frankfurt was one of the first to breed captive gorillas. The first of its female gorillas to give birth later produced female twins, the only known gorilla twin-birth.

The Frankfurt Zoo is closely associated – mainly through Dr Grzimek, who retired as Director in 1974, and who was succeeded by Dr Richard

Faust, another conservationist – with the worldwide movement to protect wildlife. Dr Grzimek's film *Serengeti Shall Not Die* was the spur to useful action not only by the Frankfurt Zoo but also by governments and societies all over the world. The Zoological Society of Frankfurt established and maintains the Serengeti Research Institute and helps in other projects aimed at the protection of rare animals. Most of this work is financed by a special fund called Help for Threatened Animals, set up by Dr Grzimek and administered by The Zoological Society of 1858.

The Frankfurt Zoo is also involved in scientific and educational work in Germany. The thirteen-volume *Animal Life Encyclopaedia* edited by Grzimek has replaced Brehm's *Tierleben* as a sourcebook on natural history. Each year the Zoo's education department arranges lectures and guided tours for more than twenty thousand schoolchildren from Frankfurt and the surrounding area.

While Frankfurt is historically the senior zoo of the southern part of Germany, two others in the region – Munich and Stuttgart – have achieved, in their comparatively short existence, international standing as major zoological exhibitions.

After several false starts due to the First World War and the economic chaos which followed, the Munich Zoo opened in 1928 in a large park well watered by streams and lakes. It was laid out on a zoogeograhical basis, with the animals grouped according to their continental origins. The latest and most impressive feature is the Polarium where polar bears, seals and penguins are shown in fine modern enclosures with large areas of water. Munich's first Director, Heinz Heck, was a member of the famous German zoo family. One of his aims was to try to breed models of the aurochs and tarpan, extinct species of cattle and horse, by selective breeding from modern domestic animals. Munich Zoo suffered damage during the Second World War, and among five African elephants killed by a single bomb was a male called Adam, the first such elephant to have been born in captivity.

The Stuttgart Zoo is situated in Wilhelma Park, named after King Wilhem of Württemberg, who built a country residence there in the mid-nineteenth century. After the First World War the Park was developed as a public botanical garden, using the classic formal gardens and ponds as the centrepiece. A palm house and houses for other tropical plants were built, and many ornamental and exotic trees were introduced. After the Second World War, during which the Wilhelma was badly damaged by bombing, a small Aquarium was added, followed by other animal buildings, and a zoo gradually took shape. A new Aquarium and Reptile House, designed to fit in with the formal layout of the botanical garden, is now the finest feature in the Zoo. It is a marvel of technology, consisting of a series of simulated water habitats to display fish and reptiles. A hillside above the formal gardens has been acquired. and well-designed animal buildings and enclosures have been built.

One of the features of German Zoos when they started in the nineteenth century was the close relationship that existed between their various directors. It was never unusual for a man who had served a term in one zoo to move to another, so carrying with him a rich body of experience. In this way German directors soon formed a small and closely-knit band – professional 'zoo-men' who, if not prominent in zoological research, were certainly that in the exposition of natural history. Happily the tradition continues to this day – to the benefit not only of all German zoos but also zoos in other parts of the world.

PHILADELPHIA

Philadelphia Zoological Garden

John Toovey

Like so many of the world's outstanding zoos, that of Philadelphia owes its origin to the initiative of a small band of enthusiasts. Their leader was Dr William Camac, a man who had been active in the city's affairs from about 1850 and who, in the course of his travels to Europe, had visited many zoological gardens. As a result he had become convinced that what both Philadelphia and America needed was a proper zoo – not the small menageries which were usually adjuncts to a museum, and to which the public was as a rule indifferent. In 1859 he succeeded in founding a zoological society, but the outbreak of Civil War two years later delayed further action for some thirteen years. Philadelphia's Zoological Society was then incorporated for the purpose of acquiring and maintaining a collection of animals 'for the instruction and recreation of the people'.

Fairmount Park, just outside the city, seemed an ideal location, and a lease of about 32 acres (13 hectares) west of the Schuylkill River was granted by the Park Commissioners. The land included 'Solitude', a miniature Georgian mansion built in 1785 by John Penn, grandson of the Quaker founder of the State of Pennsylvania and, at the time of its sale to the Park Commissioners in 1867, the last small piece of Pennsylvania in the hands of the family which had once owned the entire state. The house was used until 1972 as administrative offices for the Zoo. Four years later, when a new administration and education building was completed, it was restored to its original state and opened to visitors as a national monument.

In 1873 about 10 acres (4 hectares) of land around 'Solitude' was fenced, and an engineer, Hermann Schwarzmann, was sent to study zoo-building in Europe, where he was especially influenced by the London Zoo. At the same time Frank Thompson, an experienced animal collector, then in Australia, was appointed Superintendent and authorized to collect animals. Within a remarkably brief space of time the first zoo in the United States opened, on 1 July 1874, to a fanfare of bands and banners. Amongst the 282 exhibits were Australian and North American species, as well as lion, tiger, zebra, elephant, antelope and rhinoceros. During its early years many famous people donated animals to the Zoo: President Grant sent two curassows; the Mormon leader, Brigham Young, contributed two black bears; and Mrs Sherman even offered the cow Atlanta which had accompanied her husband, President Grant's illustrious general, on his famous march through Georgia.

With more than 200,000 visitors in the first year, the Society pushed ahead with its building programme so that as much as possible would be ready for the 1876 Centennial Exhibition Year in Fairmount Park. This brought in over 677,000 visitors, at 25 cents per adult and 10 cents per child. In those days a zoo visit was a leisurely and popular affair. Horses would be hitched in their hundreds to the gates and fences, and there was a landing stage on the Schuylkill River for such visitors as came by steamboat. Funds were also boosted by an issue of stock certificates, which were quickly taken up, enabling the Zoo to be completed within its first ten years at a cost of about $195,000.

Once, however, the novelty of the new zoo had faded, attendances plummeted and rarely exceeded 200,000, falling during the Depression of the 1930s to an all-time low of 152,000. They started to rise again after the Second World War, and from the mid-1960s onwards annual attendances have been over the million mark.

The city of Philadelphia made its first annual maintenance grant to the Zoo in 1891, and since then has also provided capital grants. These have been the mainstay of the Zoo's more recent major improvements, but operating funds have never been plentiful. Grants, private donations and bequests have helped considerably. One of the most important was a bequest of over $600,000 left in 1925 by Wilson Catherwood, a former member of the Zoo Board. Initially invested to provide income, it later became the cornerstone of the Zoo's redevelopment, and financed many of the new buildings that were constructed between 1938 and 1951. Whether the financial situation of the Philadelphia Zoo actually compared unfavourably with other American zoos is, however, open to question. The Secretary of the Smithsonian Institution, complaining to Congress in 1892 about the $17,500 allocated to run the National Zoological Park in Washington, spoke covetously of the $45,000 annual budget of Philadelphia Zoo.

The solidly constructed early buildings at Philadelphia copied the heavily ornate style of their European contemporaries, and would today be described as quaint follies. All that now survives from this early period are the fanciful 1874 cottage-style gatehouses flanking the north entrance, designed by Frank Furness, and the Antelope House by George Hewitt, both renowned Philadelphia architects of their day.

After the start of the Depression in 1929, when income and visitors were at rock bottom, buildings, paddocks and services were all virtually derelict. This was the low point of the Philadelphia Zoo's history. The standard of the animal collection had also fallen pitiably and staff morale was at its lowest. The Society's fortunes took a significant turn for the better with the election in 1926 of Dr Williams Cadwaladar as President. He started by improving the quality of the senior staff, bringing in Mr Freeman Shelly as Director and Dr Roger Conant as Curator of Reptiles with responsibility for public relations. The administration was reorganized and modernized, and reconstruction begun with the aid of the Catherwood legacy. A sensible start was made with a service building, which had not existed before and which brought together the animal food kitchens, workshops, garages, etc., greatly improving working conditions and efficiency. This was followed in 1941 by the Pachyderm House for elephants and rhinos. Its moated open enclosure caught the public's fancy and sent attendances up to over 500,000.

Once the Second World War had ended, and building materials and labour had become available, the modernization and rebuilding programme got going again. In 1948 the 1888 wing of the Reptile House was remodelled. In 1950 the 1916 Bird House was modernized and enlarged. The first completely new building was the Lion House of 1951, which was designed to house twenty-five species of large cat, with moated outdoor enclosures for lions and tigers. It cost a million dollars and, while largely financed by the city of Philadelphia, it also absorbed most of the Zoo's available resources. After that development slowed down. In 1957 a new Children's Zoo, made possible through a generous grant, replaced the Baby-Pet Zoo – built in 1938 – which had been the first children's zoo in America. Houses for small mammals were built in the early 1950s and, in contrast to the hygienically tiled Lion House, were designed as 'simulated habitat' exhibits.

It was the walk-through humming bird exhibition, attached to the Bird House, that proved to be the first of the post-war developments to attract the attention of other zoos. Opened in 1970, it was planned by John Griswold, Curator of Birds, to show tropical New World plants and birds in a typical habitat. A new Reptile House in 1972 followed the same trend, with exhibits which used as much natural planting as possible against a naturalistic background built up in glass fibre and other artificial materials.

In 1973 the habitat conception of exhibition was extended out-of-doors to a new 'wolf wood' and an African Plains exhibition which was laid out by Philadelphia architects Mirrick, Pearson, Ilvonen and Batcheler on an area previously occupied by the hoofed animals that had always been a feature of the Philadelphia Zoo. This area, which measures about 5 acres (2 hectares), is long and not too deep, so that the animals are always within sight of the public. Wherever practicable, paddock surfaces are grassed and the area is furnished with such typical African features as waterholes, a termite mound and a large artificial baobab tree. Even the animal houses and associated yards are camouflaged behind the right kind of artificial rock-work. Stage one opened in 1974 with a mixed display of zebra, giraffe, antelope, ostrich and crane, and Stage two followed a year later with white rhinoceros, springbok, blesbok and various long-legged and ground-living birds. Part of an aerial monorail constructed in 1969 runs alongside the Plains exhibition, giving a splendid panoramic view of the scene. The African Plains exhibition cost one-and-three-quarter million dollars, and Philadelphia has now raised the funds for another major exhibition – Bear Country, by the same architects, due to open in 1979.

The Zoo took over a fine, mature, wooded and landscaped garden, with the house 'Solitude' as its focal point, and on which the Society from the start spent large sums. A propagating house was built in 1875 and a palm house in 1877. An extract from *Harpers Magazine* of 1879 suggests that the money was well spent:

> In the center is a lovely fountain, and in the summer the low wall enclosing the terrace is a mass of green; it is completely covered by trailing vines. . . . The head gardener of the Society must be a man of taste and ability. This is evident everywhere throughout the grounds . . . the ribbon beds of many-tinted foliage plants and brilliant flowers are dazzling in their beauty, and nowhere show puny or sickly response to the gardener's skill . . . and finally the wealth of grand old trees, many of them gigantic in size, and the rolling surfaces of the land render the whole place naturally picturesque and beautiful.

The influence of the unnamed gardener referred to in this paragraph never faded. Many of the fine trees of today were planted in the decades which followed and, as they become affected by new construction, old age or disease, they are replaced. Many fine animal sculptures, of which the Society owns a unique collection, adorn the lawns and the displays of annuals, flowering shrubs and exotic plants. The first major acquisition, *The Dying Lioness* by Wilhelm Wolff, has stood outside the north gate of the Zoo since 1874.

The growing use of habitat displays has meant that the gardening staff have become increasingly involved in the upkeep of the animal exhibits. Special skill is needed in selecting the right vegetation – particularly where local conditions and climate do not permit the use of appropriate plants – and then in protecting it from the damaging attention of the animals. 'Natural' habitats are, in practice, realized only through continuous experiment, unremitting maintenance and hard work. The botanical collection at Philadelphia is almost as important as the zoological one.

Philadelphia, like many non-commercial zoos throughout the world, has been consistent in its attention to the quality of its animal and botanical displays. So what puts it amongst the great zoos? The answer is that it has always promoted research in veterinary care, and that its breeding and longevity records have been outstanding. When the Society was founded, one of its stated aims was 'that the scientific purposes of the Society shall be carefully regarded'. In the pattern of the Zoological Society of London, on which Philadelphia tried to model itself, Professor Henry Chapman was appointed in 1874 to the post of Prosector. His first report to his Council, for the period January 1875 to May 1876, contains some prophetic remarks on animal management:

> My experience as Prosector to the Zoological Society convinced me that, during the first six months of the existence of the Garden, the principal causes of death were three: Ist, improper food, both in quantity and quality; 2nd, effects of temperature; 3rd, ill-constructed cages, wanting in sufficient space and deficient in necessary appurtenances, such as water to bathe in, trees to climb, soil to burrow in, etc., according to the nature of the animal.

He advised that every animal should be separately studied and recommended 'placing them under conditions approximating as nearly as possible to those of nature's'.

For a short time the Council's interest in scientific management waned, but in 1901 two Professors in the Medical School of the University of Pennsylvania realized the value of studying the pathology of extant animals. One was Dr Charles Penrose, a gynaecologist and obstetrician who worked closely with the Society, and who served as its President from 1909 to his death in 1925. The other was Dr Cortland White, a pathologist, to whose Pepper Laboratory the Zoo's dead animals were sent, and where they were given accession numbers in sequence with the human autopsies, being recorded as Mr L. L. Gorilla, Mrs P. H. Antelope, Mr Red Patas, etc. The slides and tissues of the first animal autopsy, a pronghorn antelope, are still on exhibition in the Laboratory. A few years later Dr White was formally appointed Pathologist to the Society and provided with a new laboratory and animal hospital, partly financed by a grant from the city, and with much of the equipment donated by Dr Penrose, by whose name the Zoo's research laboratories are now known. Dr Penrose was a man of vision, and the comparative pathology which he pioneered is recognized today as an essential branch of medicine.

The work of the new laboratory was not confined to routine pathological studies. Different ways of improving animal housing and health were also tested, and in 1905 Dr Penrose and Dr White were joined by Dr Herbert Fox, with whom they tackled the problem of tuberculosis, which at that time was a scourge of many monkey collections. As one result of their findings, Philadelphia's primates – the monkeys and apes – have for many years been kept behind glass to avoid cross-infection between the animals and the public.

TOP OF PAGE A white eared pheasant, a native of China, at Jersey Zoo.

ABOVE Several generations of the Manchurian crane now live at Tokyo's Ueno Zoo.

OPPOSITE At Antwerp Zoo, the wisent or European bison (*above*) once nearly extinct, and the okapi (*below*), only discovered at the beginning of this century.

PREVIOUS PAGE Przewalski's horses at Prague Zoo, where the international studbook for this rare species is held.

ABOVE The margay is one of four different species of small cat kept in the Arizona-Sonora Desert Museum's man-made rock grotto.

OPPOSITE ABOVE Sumatran tiger at London's Whipsnade Zoo.

OPPOSITE BELOW Cheetah and cubs at Pretoria Zoo, where they live in near-natural conditions.

PREVIOUS PAGES A group of Somali wild asses at Basle Zoo.

LEFT A Rodrigues fruit bat, possibly the rarest bat in the world, at Jersey Zoo.

BELOW The maned wolf, a rare species from South America, at East Berlin Zoo.

Professor Chapman's admonition of 1867 that the complex disease patterns found in zoo animals, including juvenile rickets, might be caused by an inadequate diet was fully appreciated by the staff of the laboratory. Dr Herbert Ratcliffe and Dr Ellen Corson-White were appointed assistants to Dr Fox in 1929, and in the five years which followed were able to formulate diets (the 'Ratcliffe Diets' now famous throughout the zoo world) for all the mammals and birds in the collection. The composite uncooked food for primates known as monkey cake was particularly successful and is in use today virtually unchanged.

The Philadelphia Zoo has the credit for a considerable number of longevity records for captive animals. The orang-utans Guarina and Guas lived to the ages of fifty-six and fifty-seven respectively, and produced nine offspring. The chimpanzees Marianne and Sultan were thirty-three and twenty-eight years old when they died. Massa, the world's oldest captive gorilla, was forty-seven in 1978, outliving his famous thirty-three-year-old predecessor Bamboo. Among longevity records for non-primates are forty-nine years for a common echidna, thirty-eight for an Indian elephant and forty-two for a hippopotamus. An alligator snapping turtle which has lived for sixty-two years holds the record for reptiles, and boa constrictor and a griffon vulture both achieved thirty-seven years. Many other creatures – turtles, bears, emus, cockatoos – have lived for over thirty years in the Philadelphia Zoo. The Zoo also enjoys a well-deserved reputation for breeding and rearing. It can boast in addition some noteworthy 'firsts' in the United States, including the Inca tern, the trumpeter swan, greater and Chilean flamingos, fennec fox, impala, dik-dik and springbok.

Papers based on the research of the Philadelphia Zoological Society appear in various scientific journals and also in the Annual Reports of the Penrose Research Laboratory. In 1923 Dr Fox published *Disease in Captive Wild Mammals and Birds*, the first general work ever on the subject. It was based on the accumulated records of nearly six thousand autopsies, and remains a standard work of reference. Some of the staff of the Penrose Laboratory hold faculty appointments in the University. The Zoo has been associated for many years with the School of Veterinary Medicine at the University, but it is only relatively recently that it has appointed its own veterinary staff. The Zoo's veterinary work comprises medical treatment, preventive medicine and quarantine testing, pest control, and the checking of food quality. The veterinary staff also advise on the design of animal housing and exhibits. Facilities include a pharmacy, a clinical pathology laboratory, and a medical-surgical room. The latter is suitable for the treatment of small to medium-sized animals. Anything larger is handled in association with the University School.

The Society has long been concerned to encourage cooperation in breeding-loan programmes, as well as exchanges of information with other zoos. It is now involved in a cooperative venture known as 'Zoo East Survival Trust' with the Bronx and Washington Zoos. Philadelphia's influence amongst American zoos stems from this concern, as well as from it having been the first on the North American continent.

BASLE

Der Zoologische Garten Basel

Colin Rawlins

No zoo can prosper unless it is secure financially. The most illustrious of Swiss zoos, at Basle, has seldom had to worry on that score. It has enjoyed the generous support not only of the city authorities, but also of enterprises such as Ciba-Geigy, Hoffman-La Roche and Sandoz, for Basle is the home of some of the world's successful businesses, particularly in the pharmaceutical and chemical industries. The Zoo has also received large benefactions from individual citizens. Like Switzerland itself, the zoo is characterized by a stability which is the envy of most others in the world.

Der Zoologische Garten Basel, known to the people of the city by the affectionate abbreviation 'Zolli', is, like most things Swiss, neat, well-planned, attractive in display and efficient in operation. Although not a big zoo if measured by the number and range of the animals it houses, it is recognized as a model of all the good modern zoo should be.

Throughout its hundred-year history, the Basle Zoo has always been improved with new exhibits and new buildings and through the introduction of new ideas about animal management and display. Since the Second World War, it has pioneered several radical changes in animal management and exhibition, under the leadership of men like Professor Geigy, the head of the pharmaceutical company which bears his name, and the direction of Professor Dr Heini Hediger (1944–53) and his successor Dr Ernst Lang. It was accepted that a good zoo should not be simply an attractive exhibition, but also a place in which captive animals breed and prosper. The species that have been exhibited have so far as possible been selected because they were in need of protection, and no attempt has been made to keep the usual range of species normally to be found in a leading zoo. Animal management has been based on a scientific study of nutrition and behaviour and, whenever practicable, cages and enclosures have been designed to simulate a wild habitat, and with careful regard for such matters as correct temperature and humidity and suitable good surfaces.

The Basle Zoo was founded as a private non-profit-making company by a group of prominent members of the Swiss Ornithological Society whose purpose, as set out in a public appeal, was to bring back to the people of the fast-growing towns and cities of Switzerland a feeling for nature. 'As the first Swiss zoological garden', the appeal declared, it would 'above all present to the visitor as completely and as naturally as possible the splendour and beauty of our Swiss, and particularly Alpine animals; at the same time, however, the more interesting European animals, insofar as they were suitable for a zoological garden, should also be properly represented.'

At its opening in 1874, the Zoo covered 11 acres (4.5 hectares) of leased land. Ten years later more was added and, with further additions in 1929, 1934 and 1959, the total area today is about 32 acres (13 hectares). In 1934 a gift by the goldsmith Ulrich Sauter enabled the Zoo to buy some 5 acres (2 hectares) of the land it was occupying. This is now known as the Sauter Garden and, with a small piece used for service buildings, is the only land actually owned by the zoo. A series of agreements with the government, from 1891 onwards, turned over other parts of the Zoo on a free lease, the term of which was extended in 1970 to one hundred years.

The Sauter legacy was only one of many gifts which Basle Zoo has received. The 750,000 Swiss francs given by Johannes Beck in 1901 were probably the most important donation; it came at a time when the Zoo was still struggling financially. Beck was a successful businessman in the silk and textile trade, and a condition of his legacy was that entry to the Zoo should be free to the public of Basle for one evening each year, and that the occasion should be a festive one. This condition has been adhered to scrupulously up to the present day. Other generous supporters of the Zoo include the scientists and industrialists Dr Aurèle Sandoz, Dr Emile Dreyfus and Professor Rudolf Geigy.

In its first full year, 1875, 100,000 people visited the Zoo. The corresponding figure one hundred years later was more than 1,000,000. The increase in income over the same period from entrance charges of all kinds was one hundredfold – from some 32,000 to over 4,000,000 Swiss francs . . . a large part of the increase being a measure of inflation.

When it opened, the Zoo housed about 100 mammals and some 400 birds, nearly all, in accordance with the declared aim of the founders, representatives of European and particularly Alpine species. Very soon, however, this selectivity had to be abandoned. Alpine animals proved difficult to keep and mortality was high. Even more important was the reaction of the public, which soon made it clear that they also wanted to see exotic animals of the kind exhibited by many other European zoos. The Zoo was in financial trouble, and even closure was being considered. In order to attract visitors, monkeys, kangaroos, leopards and a tapir were therefore introduced. An Indian elephant, which lived for thirty years, was acquired in 1886. The first pair of lions arrived in 1890 and in the following year produced their first cub.

The number of mammals and birds increased slowly up to the end of the 1950s but then, in accordance with a new policy of specialization, began to decrease. The total of mammals and birds now stands at just over 1,300, and the number of all creatures in the Zoo, including fish, reptiles, amphibians and invertebrates, is now just above 4,000.

From its beginning the Basle Zoo has been a true garden, with trees, shrubs, grass and water providing a natural background for the animals. Buildings, roads and formal gardens have been made as discreet as possible. The visitor's first view on entering the Zoo is a landscape of meadow and water with overhanging trees. The real Zoo lies behind this green curtain.

Most of the Zoo's buildings are situated along its higher north-western side, with the service area tucked away at the western end. The animal buildings are generally well hidden, some by clever planting and others by being built underground. Moats, which are either invisible to the public because of the planting, or which are filled with water, form the barriers to many animal enclosures. A stream runs through the wooded area in the lower south-eastern parts of the Zoo.

Nearly all the main buildings have been either built or renovated during the past twenty-five years, one exception being the excellent restaurant which dates from 1935. Another fine landmark from the past is the Sealion Rock, an exhibit of the highest quality, built in 1922 by a well-known Swiss mason called Eggenschwyler. The Elephant House, with its outside arena, was completed in 1953. It was followed by a new Carnivore House in 1956, a new Ape and Monkey House in 1969, a Vivarium in 1972 and a Children's Zoo in 1977. The Ape and Monkey House represented a big advance in zoo design, with much more attention than before being given to the behavioural requirements of the animals, and with glass instead of mesh or bars being used for the large indoor cages. In the Vivarium an ascending spiral walkway, with fish exhibits on either side, is a new and attractive development. Much of the cost of these new buildings was borne by gifts from Basle citizens and industrial companies, or by grants from the canton and city governments.

The Zoo houses breeding stocks of many species which are regarded as endangered. Notable among these are the waldrapp or red-cheeked ibis, the douc langur and proboscis monkey, all the great apes (especially the gorilla), the spectacled bear, Somali wild ass, Indian rhinoceros, pygmy hippo and bongo. The breeding record for captive gorillas has been outstanding. A female gorilla, Achilla, acquired in 1948, was joined in 1953 by a young male, Stefi, from the Columbus Zoo in the United States. Achilla gave birth to a female, Goma, in September 1959, the first gorilla birth in any European zoo (one had already been born in the United States). In May 1971 Goma gave birth to the world's first captive-born second-generation gorilla. By the end of 1977 nine gorillas had been born at Basle.

The consistent breeding of the Indian rhinoceros is another of Basle's success stories. In 1951 and 1952 a pair of these animals, of which only a few had previously been exhibited in zoos, was acquired, and a birth, the first in any zoo, took place in September 1956. By the end of 1977, fifteen Indian rhinoceroses had been born at Basle, more than all others born elsewhere in captivity during that period.

No account of the breeding achievement of the Basle Zoo would be complete however without reference to the pygmy hippos, also an endangered species. The first specimens were acquired in 1932. Since then more than fifty have been born. Basle has now become a source of supply of this animal for many zoos.

Among many other species which have bred at Basle are the African elephant, the okapi and the Malayan tapir. The parents of the African elephant which was born in 1966 arrived with six other young animals just before the new Elephant House was opened. The group of eight became a popular feature of Basle, since they were regularly paraded, in a long obedient line, through the streets of the city. The children knew each of them by name.

As has been indicated, the successful breeding record of the Basle Zoo can be attributed mainly to the scientific attention which has been given to three aspects of animal management – nutrition, accommodation and behavioural requirements. From its beginnings the Zoo has maintained a close relationship with the local university, and in recent years it has also formed a scientific association with the Institute of Animal Pathology in Berne University. During the period of the presidency of the zoologist Dr Fritz Sarasin, from 1921 to 1941, the animal collections proved their value in scientific study, which they continued to do during Professor Geigy's lengthy term of office as President, from 1941 to 1972. The Zoo's educational role has not been confined to scientific studies at post-graduate level. Some 30,000 schoolchildren and students visit the Zoo each year, the former coming under the free concession which is given to children from the schools of the city and its surrounding area. Courses in zoo-biology for advanced students from the Universities of Basle, Berne and Zürich are held during the summer, and in recent years there have been courses of instruction for biology teachers from the middle and upper school classes of local schools.

In 1919 a private society, now called the Union of Friends of the Zoo, was formed primarily to help with raising money to pay for animals. Its magazine *Zolli*, which appears twice a year, provides a regular and detailed review of Zoo affairs, including descriptions of new developments, and articles on animals in the collection.

Some of the people who have helped to make Basle a great zoo have already been mentioned. Dr Beat Sarasin, an eminent member of a family that has always been connected with the institution, is now President of the Zoo's Council, as was his grandfather before him. The Basle Zoo need not fear that it will ever lack the kind of support without which a national zoo cannot hope to prosper.

SYDNEY

Taronga Zoo

Peter Olney

Taronga Zoo has a panoramic view over Sydney Harbour, its name reputedly deriving from the aboriginal word meaning 'view over the water'. Laid out as it is on the steep slope behind Athol Bay on the northern shore of the harbour, it probably has the most spectacular setting of any zoo in the world. It is well worth a visit for this alone. Scenically the best approach is by ferry across the harbour to a landing stage at the base of the hill. From there most visitors prefer to take a short bus trip to the top and then make their way downwards through the Zoo, and so back to the wharf. Seen from the harbour, the Zoo dominates the northern shore, and man-made features are relatively well screened by trees and bushes. Most of the adjacent land is either natural bushland or bushland laid out for recreation, and once inside the Zoo the visitor cannot see the residential area of Sydney to the north.

Like other Australian zoos, Taronga specializes in Australasian fauna, both from necessity and by choice. Australia has strict controls over animal imports, and comparatively few exotic animals are allowed into the country, which fortunately possesses its own unique and extensive native fauna of which most Australians, unless they visit a zoo, see only the larger and more obvious forms. Equally, because of the country's strict export controls, many Australian animals cannot be seen outside the country.

Sydney's Zoo was first sited south of the harbour in Moore Park. Here, in 1881, on a 7-acre (3-hectare) site called Billygoat Swamp, the Zoological Society of New South Wales started a zoo. This was the third public zoo in Australia, Melbourne having opened one in 1857 and Adelaide in 1878.

By 1912 the Zoo had been extended to 15 acres (6 hectares) and contained over three hundred species or subspecies of mammals, birds, reptiles, amphibia, fish and invertebrates. But since there was no further room for expansion, the New South Wales parliament had in the preceding year granted the Society a 43-acre (17-hectare) site in Ashton Park, on the north side of the harbour, an area which was later increased to 75 acres (28 hectares). Construction work began in 1912, and in 1916 the Moore Park collection was moved to what is now Taronga Zoo.

The land had been granted on condition that the Zoo would be administered by a Trust consisting of eight people appointed by the Governor of New South Wales. Although the Society continues to be associated with the Zoo, control was consequently passed in 1915 to the Taronga Zoological Park Trust. The guiding and very forceful spirit in the affairs of the Trust, almost from its beginning, was, until his retirement in 1967, Sir Edward Hallstrom. Sir Edward was a powerful and colourful man, and a well-known eccentric, with interests ranging from kerosene refrigerators (in the manufacture of which he made a fortune) to gliding. He was a great benefactor, and presided over many societies apart from the Zoo. One of his hobbies was collecting hats that had been worn by famous men – his prize possession being a wide-brimmed white stetson that had belonged to Winston Churchill. But animals in general, and Taronga Park in particular, to which he devoted a vast amount of time and donated much money, were his main interest. His name is especially linked with birds-of-paradise, in which he became interested during visits to his farms in New Guinea. There he kept a large private bird collection, as he also did in Australia, and Taronga often benefited from both of these. He was also instrumental in building up stocks of giraffes and rhinos in Australia – for he seemed to be a law unto himself when deciding what animals should be allowed into or exported from the country. On an official visit to New Guinea with Lord Mountbatten, Lord Zuckerman, then Secretary of the Zoological Society of London, was presented with an echidna. Hearing that they were in 'his' territory, Sir Edward Hallstrom flew over from Australia and promptly confiscated the animal, which he said could not be exported. Shortly after the return of the party to England, however, Sir Edward sent over to the London Zoo a pair of echidnas and also a collection of birds-of-paradise. One of his lasting contributions to Taronga was a 40-acre (16-hectare) farm which he sold to the Trust in 1961 for a token price, and from which he had provided the Zoo with free fodder for many years. Until 1976, when it became unworkable, the farm produced up to 800 tons of green fodder and 70 tons of carrots, lettuces and sweet potatoes for the Zoo animals every year.

Sir Edward had a passion for concrete which was not shared by many of his colleagues, and indeed until recently this material was a dominating feature of the Zoo. He believed in its necessity partly because of the steepness of the site which called for much terracing, but mainly because he held – contrary to other opinion – that only by providing easily cleaned surfaces could diseases and parasites be kept at bay. He was not concerned with drugs, especially those given prophylactically, which could have controlled most parasites and many diseases.

In spite of Sir Edward's firm direction and open-handed generosity, the Zoo was nonetheless in serious financial trouble by the mid-1960s, chiefly because of falling attendances. Many enclosures were either in bad repair or unsatisfactory by modern standards, and considerable redevelopment was called for. In 1966 the New South Wales parliament accordingly commissioned a report on the Zoo's affairs, and this led to drastic changes in its administration and funding. In the following year a planning committee, consisting of representatives of the Taronga Zoo Trust – but not including Sir Edward – and of the Government Architect's Office, was set up at the request of the Under-Secretary of Lands for New South Wales. Their decisions have led to major changes in the Zoo in the past few years, particularly in administration and funding. Then in 1973 the Taronga Zoological Park Trust was dissolved and replaced by the Zoological Parks Board of New South Wales, a 'statutory body with power to establish other state zoological parks, and with defined responsibilities in education and research'. Sir Edward had died by then, but he must have had the last laugh. For even if the reaction to his views about the way to run a zoo had led to the intervention of the Government in 1966, and shortly after to his resignation as Executive Director of Taronga, a law had nonetheless been passed which made him Honorary Life Director of the Zoo's Trust.

The native land mammals of Australia consists entirely of monotremes (egg-laying mammals), marsupials (young born at an early stage of development and usually kept in a belly pouch), murid rodents (rats and mice) and bats. There is also the dingo, a form of domestic dog which was introduced into the continent several thousand years ago and which became wild. It is the only wild mammal of Australia of the kind found on the other continents of the globe. Almost half of Taronga's mammal collection is Australian in origin, and the only native mammal families not represented are the marsupial moles, which are hardly ever seen even in the wild, and the insectivorous bats, which are extremely difficult to keep in captivity.

There are only two kinds of monotreme – the duck-billed platypus and the five species of echidna, both unique to Australia. They lay eggs in the way many reptiles do but, like all mammals, produce milk to suckle their young. Platypuses are difficult to keep and exhibit, and are not seen in captivity outside Australia. Taronga is one of the three Australian zoos where they are exhibited. In the wild the creature spends most of its time in burrows, emerging briefly at dawn and dusk to feed and play in the water. In accommodation which Sir Edward Hallstrom had designed at Taronga, they emerged only towards dusk and even then could rarely be seen because of the mud they stirred up when feeding. Platypuses possess the same physiological mechanisms for regulating body heat as do all warm-blooded creatures, but the range of temperature to which they are adapted is limited; they are particularly sensitive to cold. In a new Monotreme House, which was completed in 1968–9, the creatures are on public view for only two hours each day, in a special glass-fronted tank 20 feet (6 metres) long and 4 feet (1.2 metres) wide. The tank is diffusely lit from the top while the public viewing area is kept dark. The separate feeding tanks, which are not on view to the public, are reached through a series of grass-lined tunnels and nesting boxes which simulate the natural living quarters of a platypus in a river bank. The animals are fed every day with about 1½ lb (680 g) of earthworms and yabbie (freshwater crayfish). None of the platypus eggs that have been laid in the Taronga Zoo has as yet proved fertile. The only zoo that has ever bred the creature is the Healesville Sanctuary in Victoria.

The echidna is not as difficult to keep in captivity as the platypus, and specimens are always to be seen at Taronga, which is one of only three zoos in the world where eggs that have been laid have hatched. The one egg which the female produces is kept in its pouch, which is rudimentary compared to that of the kangaroo. When the young breaks out it remains in the pouch where it suckles. Unfortunately not one echidna which has been hatched in captivity has survived. In the wild they live on ants and termites, but they adapt readily to more mundane diets in captivity.

Another Australian ant-eating mammal is the numbat, a rare marsupial now found only in the south-west of western Australia. It feeds exclusively on termites, and Taronga is the only zoo that has kept the numbat in captivity for any length of time – one for nearly eight years. Two litters have been born, but no young have been reared. The breeding pair has now died.

Yet another extreme rarity in Taronga is the mountain pygmy possum (*Burramys parvus*) which until 1966 was known only from fossils. In that year a live specimen was caught in a ski-lodge in the mountains of Victoria, and the Wildlife Division of the Commonwealth Scientific and Industrial Research Organization then started a captive colony of pygmy possums in Canberra. In 1971–2 a second colony was established at Taronga. Pygmy possums are fed on oily seeds and mealworms, and more than ten have been born and reared in the Zoo.

Needless to say, the koala bear, known in toy form to tens of millions of children almost as well as the 'teddy-bear', is a constant even if shy exhibit in the Zoo. It lives in the forests of eastern Australia, and now has a very restricted and fragmented distribution, even if it is common in a few places. Koalas feed almost exclusively on the leaves of a few species of eucalyptus (gum) tree. The koala enclosure at Taronga is a model of how arboreal animals should be exhibited in captivity. When kept in trees, which seems the natural way to keep them, koalas are difficult to see. The problem has been solved in Taronga by keeping them in an enclosure furnished with dead eucalyptus trees that are dressed daily with fresh leaves. A helical ramp built of wood allows the animals to be seen at all levels. At the top the ramp straightens out, and a semi-circular shelter protects both animals and visitors from the prevailing south-westerly winds. Several koalas have been born and reared in this simple but functional exhibit. Some non-Australian zoos have tried to keep these animals, but unsuccessfully, and today they are to be seen outside Australia only in the collection at San Diego.

Kangaroos and wallabies are Australia's best-known marsupials. They are classified in some fifty species, of which Taronga keeps and breeds about ten, among them the very rare brush-tailed bettong. Another rarity is the parma wallaby, which has bred so well at Taronga that twelve surplus animals were let loose in 1971–2 in a sanctuary on Lake Macquarie in New South Wales.

A house for nocturnal marsupials was opened in 1972, and this made it possible to view some twenty species which otherwise are rarely seen. The building is sunk into the ground, and its roof is landscaped as a rest area for the public. Temperature and simulated moonlight are controlled separately for most of the cages. There is also a large area not open to the public where breeding colonies of small marsupials are maintained.

Taronga owns a representative collection of mammals from continents other than Australia, including almost thirty ungulate species, among them Himalayan thar, hippopotamus and pygmy hippopotamus, black rhinoceros, sitatunga, and Père David's deer. It also maintains a good collection of monkeys and apes, its colony of chimpanzees probably being the largest in any zoo in the world. The presence of these exotic species is something of a feat in view of Australia's strictly enforced import controls, the purpose of which is to prevent exotic diseases or potential pests being introduced from abroad. The country has never suffered from rabies and only rarely from ungulate diseases. To protect its cattle and sheep industries, the import of hoofed animals was banned in 1968. Another reason why the Australians are wary of the introduction of non-Australian species of animal is that exotic creatures like the rabbit had at one time spread until they became costly and widespread pests. The well-intentioned activities of acclimatization societies in the 1800s had led to many species, for example hares, deer, foxes and cats, getting loose and then spreading in the wild. Today the import of small mustelids, such as ferrets, or rodents, such as hamsters, and certain other groups is virtually prohibited. Primates and most carnivores are allowed in, but must undergo a period of quarantine in the Taronga or the Melbourne Zoo. Cooperation between Australian zoos has been encouraged by the fact that all zoo mammals live virtually in perpetual quarantine, and can be moved only from one to another designated zoo.

The import of birds is controlled even more strictly. Ornithosis and Newcastle disease, which are a scourge of European poultry, are not significant problems in Australia. In order to prevent the introduction of any disease, there has been a total ban on any imports of birds for many years. Since birds do not breed as well as mammals in captivity, it is a tribute to the good management of the Taronga authorities that about one-sixth of the birds they exhibit are exotic species whose forebears were imported before controls were introduced.

Australia has a large and fascinating avifauna of its own, with some eighteen endemic families or subfamilies and many unusual species. Over 60 of the 160 families of birds in the world are represented at Taronga. In recent years the zoo has started to display its birds according to their habits or habitats, with exhibits of birds characteristic of Australian heaths, marshes, treetops and desserts. There are also interesting waterfowl and wader exhibits.

The largest single habitat display is the walk-through rain-forest aviary which was opened in 1972. It contains about 30 species typical of those found in Australian coastal forests, which now occur only in small pockets along Australia's east coast, where they consist either of a mixture of tree species, or predominantly of eucalyptus. The planting of the aviary was successfully designed to simulate particular characteristics of both these forest types, while birds have been chosen to represent different feeding habits and different ways of life. Like the Sydney Opera House across the harbour, the design of the aviary was something of an engineering feat but, unlike the former, it cost relatively little to build and is entirely functional. Tensioned stainless steel rods and wires support two inclined parabolic arches, the frame being covered with wire mesh. There are no protruding struts. Birds as small as Australian wrens can be kept in, while rats, sparrows and other scourges of aviaries cannot get in. Visitors enter the aviary through revolving doors from which they move on to an elevated walkway made of timber.

Birds that live on the forest floor are represented by several species, including the superb lyrebird, which is the more southern species of this extraordinary avian family. Lyrebirds are shy and are rarely seen in the wild, nor are they exhibited in captivity outside Australia. They are excellent mimics not only of other bird calls but also of man-made sounds, such as an engine or a saw going through wood. The male courtship display is one of the most spectacular of the bird world. The tiny insectivorous scrubwren is also kept in the aviary. But unlike its close relatives, the thornbills, which mostly live in the 'canopy' or highest levels of drier eucalyptus forests, and which are exhibited in another part of the Zoo, the scrubwren seldom moves more than 3 feet (1 metre) above ground level.

Birds that subsist on fruit are represented by several species of fruit pigeon, figbirds (*Oriolidae*) and bowerbirds, which are also given to eating leaves. Pigeons and figbirds tend to live mainly in the crowns of the trees, but the bowerbird, of which there are nine Australian species of this mainly Papuasian family, are to be seen at all levels. Bowerbirds derive their name from the fact that the male builds a 'bower' from twigs and other available material into which he attracts the female for mating. The entrance and the area surrounding the bower are decorated with brightly coloured objects which the male bird may pick up at considerable distances from its display area. Coloured bus-tickets and bits of coloured glass are known to have been used. In Taronga's rain-forest aviary a satin bowerbird male has obligingly built his bower within 3 feet (1 metre) of the public walkway. The bower is tended and protected by the male throughout most of the year, and may be used year after year, constantly growing larger as it is reconstructed and augmented.

Nectar-feeding birds are represented in the aviary by several species of honeyeaters, although some of these are more given to the eating of fruit and insects than their name and brush-tipped tongue would suggest. Honeyeaters are among the most numerous and characteristic birds of Australian wooded country, particularly in eucalyptus woodland. They play an important role in helping to pollinate the plant species on which they feed.

Most of the fifty species of parrot which are native to Australia can be seen at Taronga. Some of the small nectar- and pollen-feeding lorikeets are kept in the rain-forest aviary. The majority of the parrots are housed in other parts of the Zoo. One particularly interesting group are the parrots of the genus *Psephotus*, of which there are five species, one of them, the paradise parrot, now extinct. Three of the remaining four nest in trees and are still fairly common, the last, the golden-shouldered and its subspecies, the hooded, nest in termite mounds and are on the verge of extinction. Because of their beautiful plumage, they are much sought after by aviculturists, and large numbers have been smuggled out of Australia to be sold at high prices to fanciers abroad. Taronga keeps and breeds the golden-shouldered and the hooded parrots, which can be seen in only two zoos outside Australia.

Another Taronga bird speciality is the owlet nightjar. This belongs to a little-known Australasian family of birds, the *Aegothelidae*, and has usually failed to survive for more than a few days in captivity. The bird is nocturnal and insectivorous and, like the nightjar, either takes insects in flight or, like the frogmouth, from the ground. But unlike both these birds, the owlet nightjar roosts and nests in hollow trees. In 1969 a young pair which had been abandoned were rescued and presented to the Taronga Zoo. The birds learned to accept a mixture of meat, which was supplemented by insects caught in a trap placed in the cage. This ingenious technique was duly rewarded in 1970 by appearance of three fully-fledged young: the only time owlet nightjars have bred in captivity. Taronga is also one of very few zoos in the world that has bred kiwis, flightless nocturnal birds which are native to New Zealand. The last specimen in Taronga died only recently, and because of the ban on the import of exotic animals, is unlikely to be replaced.

Taronga is now redeveloping its Reptile House to provide better and more naturalistic surroundings for the animals. Representatives of more than ten reptile families are kept, and although import controls are not as strict for reptiles as for birds and mammals, the emphasis is again on the large and distinctive Australian reptile fauna. A few species from abroad are exhibited, including the rare New Zealand lizard-like tuatara, a living relic of a group of otherwise extinct reptiles which flourished some 150 million years ago. The tuatara is of interest to zoologists because it is unique among living vertebrates in having what is called a third or pineal eye on the top of its head.

Of greater interest to the general public is the frilled lizard of northern Australia, which has an enormous brightly coloured ruff that becomes erect in threat displays. Despite its size and often fearsome appearance, this creature, like the skink (a lizard-like reptile), is an insect-eater, harmless to man. Several species of both these kinds of reptile have bred in Taronga, which also exhibits several large, predatory Australian monitor lizards, as well as the Indonesian komodo dragon, the largest of all living lizards and seldom seen in captivity.

Australia has the dubious distinction of being the home of more venomous than non-venomous snakes. Most belong to the *Elapidae*, the family which also includes the cobras and mambas. Taronga exhibits all the common Australian species, including the tiger snake and the taipan, whose venom is more poisonous than that of any other snake.

The Zoo's Aquarium, which was built in 1927, is the largest in Australia. Although well-sited from the point of view of the availability of sea-water, two of its floors have had to be closed to the public. A new Aquarium has to be built, but in the meantime the remainder of the old one has been completely remodelled to display animals in accordance with their natural habitats. The exhibits include common fish of the Great Barrier Reef, fish found in Sydney Harbour, mudskippers in a swamp of live mangroves, and several invertebrate displays.

Taronga Zoo has a modern veterinary and quarantine complex, which includes a hospital, isolation quarters, and a diagnostic and research laboratory. It has also opened a drive-through sister zoo, near Dubbo, a rapidly growing town about 300 miles north-west of Sydney.

The Western Plains Zoo, as it is called, covers about 740 acres (300 hectares) and its exhibits are arranged according to their continents of origin. The Zoo specializes in large kangaroos, wallabies, herd ungulates and large long-legged flightless birds such as emus. There are lakes which attract wild waterbirds, as well as being the home of several pinioned exotic species of duck.

Western Plains is designed so that visitors can drive past the exhibits and can alight with safety whenever they wish, or drive to focal points, from which they can walk along trails frequented by free-roaming animals. In its first year the Zoo had more than 170,000 visitors, and it looks like becoming as big a success as Taronga, which now attracts nearly a million visitors a year.

When it started, reports of Taronga Zoo's activities appeared in the *Annual Reports* and *Proceedings* of the Royal Zoological Society of New South Wales. In 1968 the Zoo started to publish its own *Annual Report*, and in 1972 the Parks Board, which now runs the Zoo, began to publish a quarterly magazine called *Koolewong*. In addition Taronga Zoo cooperates with other Australian and New Zealand zoos in producing a periodical called *Bulletin of Zoo Management*. Like most other civic zoos, Taronga also has an educational programme for schools, and a building complex is now being constructed for this. The Zoo Education Service is under the state government's Education Department which seconds teachers to the Zoo for two-year periods. Most of the running costs of the Zoo are covered by gate-money. The New South Wales government provides for the Zoo's capital development.

Australian zoos could be regarded as fortunately endowed because they are sited in a vast continent with a unique fauna of its own. Some, however, might think that they are penalized because import controls make it difficult for them to exhibit creatures from other parts of the world. Sad though this may be, it is also a blessing in disguise. Australia has learnt at great cost what happens when extraneous diseases are brought in by exotic creatures and when the latter – like the rabbit – multiply until they become expensive pests. There are lessons here for all countries, and lessons too in the way the Australians have used legislation to cope with the problem and to benefit their zoos.

TOKYO

Ueno Zoological Gardens

Colin Rawlins

It is a remarkable fact that every year nearly two-thirds of Japan's population of over 100 million pay a visit to one of the country's zoos or aquaria. This is almost twice the ratio for all visitors, including tourists, to British zoos, and at least one-fifth more than the corresponding United States figure. Relative to the size of its population, Japan probably has more aquaria and zoos than any other country.

Its main zoo, the Ueno Zoological Gardens in Tokyo, receives eight million visitors a year – more than any other zoo in the world, except perhaps those in Peking and Mexico City. Founded in the late nineteenth century, it is by far the oldest zoo in Japan. Only one other, Kyoto, approaches it in age. Most other Japanese zoos were established after the Second World War, in a period when the number of aquaria was also multiplying fast. As Japanese society became prosperous and westernized, the Tokyo Zoo developed in line with all its modern counterparts in other countries. It is now world-renowned, reflecting the growth and success of Japan and the changes in the interests and habits of its people. It houses more mammals, birds, reptiles and amphibians than any Japanese zoo, while the number of its fish is exceeded by only a very few Japanese aquaria – all specialist institutions which are not part of a general zoo.

The Zoo is situated in Ueno Park, near the main university and the University of Arts. It is also close to most of Tokyo's leading scientific and cultural exhibitions, including the National Science Museum. It is thus part of the cultural heart of the vast conurbation of Greater Tokyo.

The Ueno Gardens were opened on 20 March 1882, as an adjunct of the main Tokyo museum which, when it started, was administered by the Department of Agriculture and Commerce. The Gardens then covered just over 5 acres (2 hectares) and contained few animals, most of them indigenous to the Japanese Islands. Four years later the administration of the Gardens was transferred to the Imperial Household Agency, where it remained for the next thirty-eight years. During this period there was some expansion both in area and in the number of kinds of animals kept, including the addition of such exotic species as elephants, lions and polar bears. To mark the marriage in 1924 of the Crown Prince, now the Emperor Hirohito (himself a distinguished marine biologist and a member of the small but very distinguished band of Honorary Fellows of the Zoological Society of London), control was transferred to the city government of Tokyo, in whose charge it remains.

Tokyo was devastated during the Second World War. The Zoo suffered severely, and in 1946 animal numbers had been reduced to some three hundred, or about a fifth of the pre-war stock. With typical Japanese industry and efficiency, the Zoo was then rebuilt, and within ten years the most important buildings had been reconstructed or replaced, and more animals were on exhibition than ever before. Because the basic zoo facilities had to be provided as quickly as possible, design tended to follow pre-war ideas of zoo architecture. When, in the 1950s and '60s, new ideas began to surface in the 'zoo world', the Japanese authorities became conscious of the need to modernize the Zoo's facilities. A long-term development plan was then devised and formally completed in 1970. It is now being put into effect.

Ueno has now grown from its original size of 5 acres (2 hectares) to 32 acres (13 hectares), which are divided by a public road into two gardens of equal size, the two being linked by the Aesop Bridge and by a monorail. In the eastern section, near the main entrance, in a beautiful and traditional Japanese garden of trees and flowers, is a fine pagoda which overlooks the less oriental scene of the main zoo with its western-style architecture. The grounds are rich with trees, particularly the ubiquitous flowering cherry. Flowers and shrubs surround all buildings and animal enclosures. In the older eastern section are the main entrance, offices and main services, while the western section, added after the Second World War, contains the new Aquarium, the Children's Zoo and, perhaps most important of all, the Shinobazu Lake.

This lake, which covers some 3 acres (1 hectare), is in the south-western corner of the Zoo, and has from time immemorial been a staging post for migratory ducks and geese. Sadly, however, the number of birds 'stopping over' began to decline after the war, probably because of pollution and disturbance resulting from the rapid growth of the city. In 1958 the Zoo authorities began to provide food for the birds on a regular basis, and since then their numbers have increased dramatically. Today at least 5,000 can be seen at any one time during the winter. Nesting cormorants are another popular feature on one of the islands. The colony, which started with only captive stock, was increased by wild birds and is now some 200 strong. About 150 young are hatched each year.

The post-war history of the Tokyo Zoo was marked by three major events – the creation of a satellite park in 1958, the opening of a new Aquarium in 1964, and the gift of a pair of giant pandas in 1972.

The establishment of a satellite zoo at Tama Park became necessary because no additional space could be provided in the central part of Tokyo to allow for the redevelopment of the old zoo in accordance with new ideas about conservation and display. In 1955 an area of woodland about 68 acres (27 hectares) in extent and outside the main city, was consequently donated by the State to the metropolitan government for this expansion. The first installations, two waterfowl ponds and an incubation house, were opened in the following year. The layout of Tama Park was based on the concept of exhibiting according to zoo-geographical regions, with particular emphasis on Africa and Asia. The Asian area had been completed, and just over 550 animals, with indigenous fauna well represented, were on exhibition by the official opening day on 5 May 1958, when a quarter of a million visitors came to the Park. Tama flourished, and in 1960 another 37 acres (15 hectares) were added on the north side to provide for the development of the African area. Today the Park covers some 125 acres (50 hectares).

The opening of the new Aquarium celebrated the eightieth anniversary of the founding of the Zoo. The building, which is in the modern Japanese style, overlooks the Shinobazu Lake, and is the largest and finest of the Zoo's new developments. The main collection of reptiles and amphibians occupies the top floor, and fish and marine invertebrates are exhibited on the two floors below. There are two large areas for special exhibitions, and a vast marine tank reaches from the ground floor into the first. Its glass front, possibly the largest aquarium window in the world, is some 50 feet (16 metres) long and 7 feet (2 metres) high. The comprehensive collections include as special features goldfish and jellyfish.

The Zoo's most famous animals are undoubtedly the two giant pandas, Lanlan, a female born in 1968, and Kangkang, a male born in 1970, which were presented to Japan by the people of China in October 1972. They are housed in a specially designed building consisting of three indoor dens, a service area and a small clinic, which was officially opened on 7 May 1973. The dens are fronted by glass panels, and there is also a large grassed enclosure outside, along one side of which is a wide covered passage. This arrangement ensures that everyone moving through the building will be able to have a close look at both animals.

Their progress has been followed with professional care by the three hundred staff, and with fervour by the Japanese people. The animals mated for the first time in the early summer of 1977 and a birth was confidently expected in the autumn. But to the great disappointment of all concerned, there was no baby panda, and tragically in 1979 the female died.

The Zoo's hippos were moved in 1957 to more open and more satisfactory quarters near the Shinobazu Lake, in a part of the Gardens set aside as the African Plains. Their original house was then refashioned to accommodate apes. An Elephant House, very similar to the Elephant and Rhino Pavilion of the London Zoo, was opened in 1968. A hospital and quarantine building was completed in 1973, and new houses for large and small cats in 1974.

The Children's Zoo, after occupying three different sites since 1958, is now in the western section, near the Zoo's monorail station. It covers 2 acres (0.8 hectares), an unusually large area for a children's zoo, and is an important feature of the Gardens. It contains many domestic as well as exotic animals, and an area is set aside where children can handle the creatures. There are regular teaching sessions for young visitors as well as for organized parties of adults. Demonstrations in the care and handling of popular zoo animals are given in an outdoor theatre called the Sunflower Stage.

The appearance of Tama Park, with its wooded hillsides and steep valleys, is very different from the urban aspect of Ueno. It includes two special features of considerable interest. The first, the Insectarium, is a unique zoological exhibition. Occupying 4 acres (1.6 hectares), it comprises a central building and four adjoining 'farms' where insects are bred. The first small insect exhibit was opened in 1961; the large permanent building dates from 1969. Breeding 'farms' for butterflies, grasshoppers, fire-flies and dragonflies have been added at regular intervals, and visitors can walk through them as through a free-flight aviary. For reasons of disease control, live insects may not be imported into Japan, so the collection, with some exceptions such as migratory locusts, is composed mainly of indigenous species. A fine display of exotic insects is, however, shown as a mounted exhibit in the Exhibition Room of the central building. In another part of the building is a collection of insectivorous, nocturnal and crepuscular animals such as bushbabies, bats and owls. A zoologically less important but popular feature of Tama Park is the Lion Garden, where a group of lions live in a vast walled enclosure through which a bus makes a circular tour.

Ueno houses nearly 600 mammals belonging to 100 species, over 1,000 birds belonging to 250 species, and some 750 reptiles and amphibians, with more than 10,000 fish in the Aquarium. At Tama there are about 500 mammals belonging to 70 species, and 1,000 birds belonging to 130 species, as well as the vast collection of insects, in excess of 50,000. Not surprisingly, both collections display a notable array of Japanese and East Asian animals. Breeding groups of Asiatic cranes, including the Japanese or Manchurian Crane, *Grus japonensis*, and the white-naped crane, *Grus vipio*, have been kept for several 'crane generations' at Ueno. One white-naped crane has been in the Zoo for more than thirty years. Tama Park keeps the international studbook for the Manchurian crane, as it does for the Japanese serow, a rare species of mountain goat which, apart from Japanese Zoos, can be seen only in Peking, Los Angeles and San Diego. The well-known Japanese macaque has been bred regularly at both Tama Park and Ueno, and there are normally groups of about fifty animals in both collections, the one at Ueno having been started in 1950. Several giant salamanders, another rare indigenous species, are also kept in the Aquarium at Ueno.

Many species of pheasant have been successfully bred at both Tama Park and Ueno. From the point of view of conservation, the most important are cheer, Elliot's, mikado and Palawan peacock pheasants. At Ueno there is an excellent collection of penguins, including king, black-footed and Humboldt's, all of which breed regularly.

All the great apes are represented at Ueno. There is also a small but valuable collection of rare species of wallaby, three of which, the Cape York pademelon, the sandy wallaby, and the white-throated or parma wallaby, breed regularly in the Zoo.

Species worthy of mention at Tama Park are the Indian rhinoceros and scimitar-horned oryx; there is a large herd of the latter in the African Savannah enclosure, nearly all of which were born in the Zoo. The same enclosure also contains a large herd of giraffe, which have produced over forty young since the opening of the Park in 1958.

Tokyo clearly occupies a proud place in the world of zoos. Well-qualified staff manage the animal collections as well as the public education programme. There is an excellent new animal hospital as well as quarantine quarters. Conservation programmes have been prepared in cooperation with other Japanese zoos for several Japanese endangered species, both in the wild and in captivity. Research is carried out into problems of captive animal management. In 1977, at Ueno, a white-naped crane was hatched after artificial insemination of a female with semen collected from a male bird at Tama Park. The two Tokyo zoos also produce for the Tokyo Zoological Park Society a monthly magazine called *Animals and Zoos*.

Distinguished Japanese scientists, including the Emperor himself, have guided and influenced the development of the Tokyo Zoo, particularly in recent years. Dr Tadamichi Koga, Director for most of the post-war period, has played a particularly valuable part. His successors at Ueno, all of whom also served first at Tama Park, were Directors Hayashi, Imaizumi, Asano, Ishiuchi, and now, Asakura, a world authority on the crane family.

The number of people visiting the two Tokyo zoos, often more than 100,000 in a day at Ueno, is a benefit which most other major zoos in the world do not enjoy and a problem with which they do not have to contend. But the orderliness of the Japanese people does much to alleviate the problem of crowd-management, a problem which most other zoos would no doubt welcome since its consequence is a healthy financial situation.

WASHINGTON

National Zoological Park and Smithsonian Institution

John Toovey

WASHINGTON

In 1889 the US Congress appointed a Board of Commissioners to plan the establishment of a zoological park in the District of Columbia. In the following year Sydney P. Langley, then Secretary of the Smithsonian Institution of Washington and a member of the Commission, reported on the progress that had been made. His speech to Congress included a warning:

> I am assured by most competent naturalists that few, if any, of those not particularly devoted to the study of American animals realize that changes have already occurred or are on the point of taking place in our own characteristic fauna compared with which the disappearance from it of the mammoth was insignificant ... is it sufficiently realized that, once extinct, no expenditure of treasure can restore what can even today be preserved by prompt action of a very simple and definite kind? It is such considerations as these that have induced me to ask the earnest attention of the Regents, of Congress, and of the country to the immediate necessity for action....

The appeal was successful, and funds with which to proceed with the park were authorized. Interest in the venture had already been aroused by William Hornaday, who at the time was Curator of the Department of Living Animals at the Smithsonian Museum. Three years before, when Chief Taxidermist to the Museum, Hornaday had made a much publicized tour of the United States in order to collect live specimens of the native fauna as models for study by the taxidermists. The animals which he brought back were kept in sheds close to the Museum, and an enthusiastic public immediately saw the need for a proper zoo. A site was chosen in 1889 at Rock Creek, 2 miles (3.2 kilometres) from the White House, then an undeveloped area on the outskirts of Washington.

The concern of the Smithsonian with problems of conservation was also a reason for its early association with the Yellowstone National Park. From it the new National Zoo obtained bison, grizzly and cinnamon bear, mule deer, pronghorn antelope, elk and beaver – all representative of the country's natural fauna. But the call for conservation which Langley had made so eloquently, and repeated again to Congress in 1901, went largely unheeded. Moreover, with the passage of time the emphasis of the collection at the National Zoo gradually changed as more and more creatures were acquired from far-distant lands. Indeed, after 1906 native animals were no longer listed separately in the annual census of the collection. Not until the late 1950s was the issue of conserving the national fauna of the country again urged publicly when Dr Reed, the present Director of the Zoo, took up the cause afresh, the Zoo being rewarded in the following year by the gift of fourteen reindeer and a caribou from the new state of Alaska. Since then the Zoo has adopted as its symbol the 'Vanishing Animal'.

For decades most of the animals in the Zoo had been donated as gifts, sometimes by foreign government, but usually by private individuals, many of whom had been on expeditions. More often than not the offerings were single animals, and frequently adequate accommodation was not available. But as a rule this did not prevent their acceptance. Washington, like most other zoos of the day, and like some still, was a consumer rather than a protector of wild animals. Following the death during the previous year of many animals, including a group of Cuban flamingos, Frank Baker, the then Superintendent, wrote in the Zoo's *Annual Report* for 1901: '. . . it would seem wiser to plan and erect the structures and afterwards procure the animals for them'. Only occasionally were gifts refused. Before moving in 1890 to Rock Creek, Mr Hornaday had to decline the gift of eighteen American bison from the Hon. W. F. Cody ('Buffalo Bill'), and also a tempting offer from a W. Conklin of camel, aoudad, lion, ibex, black leopard and ostrich. Just as it was about to move, the Zoo was unexpectedly offered two elephants, which it did not refuse, and for which temporary housing had to be provided.

In its earlier years the Zoo also accepted many animals on a loan basis, with the proviso that any young which might be born became Zoo property. One frequent 'lender' was the Barnum and Bailey Circus, the purchaser in 1882 of the London Zoo's famous elephant, Jumbo – a name which has since become an adjective synonymous in many languages with 'big' or 'oversize'. In 1893 the Circus deposited a tiger and a zebu; in 1899 two lions, two zebus and an axis deer; in 1904, in conjunction with 'Buffalo Bill' himself, twelve American bison; and in 1934, a fifty-one-year-old elephant named Babe.

Among the many gifts which the young Zoo received were a lion presented to President Theodore Roosevelt in 1904 by King Menelik of Abyssinia, and another lion, two zebra and an oryx which were sent by Ras Makonnen, Governor of Abyssinia's Harrar Province. Other gifts were sent by US consuls in foreign posts, the US consul in New South Wales sending as many as 140 varied specimens in 1903. In 1918 New Zealand sent a collection of birds, including six kea, and at about the same time the Zoo received from the Duke of Bedford, then President of the Zoological Society of London, a gift of four eland, four Kashmir sheep and four Manchurian stags, a species which was believed to be related to the American elk. These animals came from the Duke's own collection at Woburn. A few years later some early 'Friends of the Zoo' raised $5,000 to buy two young Sumatran elephants. In 1924 the acquisition of a Baird's tapir from Central America enabled the Zoo to show three of the four species of tapir at one time, the other two being Malayan and Brazilian.

The unplanned acquisition of animals presented many problems. On his appointment in 1925 as Director, Dr William Mann complained that '. . . of permanent structures the National Zoological Park has a primate house, a house to contain one elephant, and one end of a carnivore house'. But he was still unable to stem the flow of new arrivals. In 1926 alone there were 150 gifts, including six rhea from South America and many birds and snakes from Sumatra.

From the start, many governmental agencies cooperated with the Washington Zoo. In 1888 the Fish and Wildlife Service of the federal government had lent its Fish Commission Box Car No. 1 to Mr Hornaday to transport by rail the animals which he had acquired during his collecting tour. The Zoo's relationship with the armed forces, starting in 1899 with Secretary Langley's request to the Secretaries of State of War and of the Navy that their officers bring back live specimens from foreign postings and tours, has proved most fruitful. The services have been a never-ending source of animals and, on occasion, particularly from their Antarctic stations, have provided some prize exhibits. Both during and after the Second World War the number of 'deposited' animals increased steadily, one interesting source being governmental agencies which were using exotic animals increasingly in various research projects. The most famous acquisition from this source was the astronaut chimpanzee Ham, an animal which was retired by the US Air Force in 1963 after a rocket flight that had taken him 155 miles (250 kilometres) above the earth.

The seemingly endless list of gifts still goes on. Particularly notable among recent acquisitions have been the 'white' tiger Mohini from India, a pair of Komodo dragons from Indonesia, four kiwis from New Zealand, and a pair of giant pandas – Hsing-Hsing and Ling-Ling – given in 1972 by the people of the People's Republic of China to the people of the United States. Washington has also regularly purchased and exchanged animals with many zoos, sometimes making reciprocal gifts such as the pair of musk oxen which were presented to the Peking Zoo, and a collection of twenty-four animals to the new zoo at Delhi, India.

Expeditions have also been a profitable source of animals, even if in the early days the results were frequently not all that had been intended. Transport was difficult and journeys long, and many animals died. Sometimes the survivors brought disease with them. This was the unfortunate outcome of Peary's Arctic expedition in 1897, when the Eskimo dogs which he deposited at the Zoo transmitted distemper to the Zoo's wolves, foxes and dogs. The 1909 expedition made to Africa by Theodore Roosevelt, and backed by the Smithsonian, was on a vast scale, but despite its supposedly scientific intent, it revealed little respect for the concept of conservation. More than $50,000 had been raised by public subscription for the expedition, the purpose of which was to obtain dead specimens for the natural history section of the Smithsonian and live ones for the Zoo. Whatever else, it was certainly a notable hunting safari. The skins of 10,897 birds, mammals and reptiles were collected and, with 2,500 sheets of plants to deal with as well, the Museum was nearly brought to a standstill. The Zoo's booty included lions, cheetahs, leopards, Grant's gazelles, zebras, elands, hartebeestes, waterbuck, warthog and crested rat. The 1917 expedition to Africa, which was partnered by the New York and Philadelphia Zoos, was somewhat less bloodthirsty. In 1920 and 1922 the zoo also received important collections of South American animals.

Despite his appointment as Director of the Zoo in 1925, Dr Mann, who had just returned from a collecting trip to the Amazon Basin, continued to go on expeditions throughout his thirty-one years in the post. In 1926 he was a member of the Smithsonian-Chrysler expedition to Tanganyika to 'remedy the African animal gap' – an historical note which makes one wonder what had happened to all the animals collected on previous expeditions to Africa.

The 1926 expedition was considered to have been extremely successful, not only because it yielded more than a thousand live specimens, but also because enough money was left over to buy a gorilla. This safari was one of the first to be recorded on film, a cameraman from the Pathé Review having accompanied the party. The public interest that was generated was considerable, no doubt aided by fifty public lectures given by Dr Mann. The result was that the number of visitors to the Zoo rose in 1927 to nearly three million. In 1937 Dr Mann was instrumental in setting up the National Geographic-Smithsonian East Indies Expedition. This roamed all the way from Sumatra to the Sudan, and again brought back hundreds of animals. Dr Mann's last tour, the Smithsonian-Firestone expedition to Liberia, took place in 1940. In the same year Malcolm Davis, a Principal Keeper at the Zoo, sailed to the Antarctic with Admiral Byrd on the first of several expeditions, and brought back a number of mammals, including crab-eating seal, and among birds the booby and a group of Adelie penguin.

There is little about any of these expeditions which would now be regarded as scientific. Expeditions today tend to be mounted for much more specific reasons. In 1959 J. Lear Grimmer studied and then brought back a group of hoatzin, an extraordinarily interesting and little-known bird from British Guiana. In 1966 Dr John Eisenberg, then newly appointed as Resident Scientist at the Zoo, went to Madagascar to study the rare tenrecs, and returned with some species which are unique to the island, including the odd-sounding fanaloka, fossa, and vontsirafotsy.

Unlike the national zoo of the United Kingdom, or that of France, the Washington Zoo did not in fact develop under what could be called serious or close scientific sponsorship. Professor Langley, its main founder, was an aeronautical pioneer (he was so early on the scene that he called his flying machines 'aerodromes') as well as being Secretary of the Smithsonian Institution. Mr Hornaday, his assistant, was a field naturalist. And even though Congress had passed an act in 1889 establishing the Park, it showed so little interest that, apart from a small annual budget, it failed to provide any funds other than a half of what was needed to buy a site for the Zoo at Rock Creek. The District of Columbia, then a poor municipal authority, had to pay the other half, even though having a zoo was not its idea.

The annual contribution from the federal government had to be authorized each year by Congress. In 1891 the executive's estimate of $101,350 was nearly halved, with a parsimonious attitude which Congress maintained for more than thirty years. In effect what Congress had done was direct that a major zoo be established for the nation at a cost to the Treasury of only $25,000 a year, while at the same time imposing on the ratepayers of the District of Columbia a bigger burden for its maintenance. It does not seem surprising now that in the circumstances the District of Columbia started to urge that what Washington needed was a conventional zoological garden, one which would attract visitors rather than the conservation park that had been envisaged by Secretary Langley, and which the 1889 Act stated was designed for 'the advancement of science and the instruction and recreation of the people' who, it was ordained, were to enter the Zoo without payment. The Washington Director could well look with envy on the far more generous budget of its sister institution in Philadelphia.

The little money that was first put at its disposal of the Zoo was not used for the animals, but for the construction of roads and for building and then widening a bridge across Rock Creek. Three years after the Zoo opened money was so short that Congress had to pass a statute just to pay a Mr Melville Lindsay the sum of $38 for supplying rubber boots to the keepers. Nothing changed until 1902, when the fact that almost a million people had visited the Zoo in the previous year at last made Congress realize that there was a problem that had to be faced. The sum of $85,000 was then authorized as the annual federal

TOP OF PAGE Pelicans at East Berlin Zoo.

ABOVE Some of the many species of wildfowl at the Wildfowl Trust, Slimbridge.

TOP OF PAGE The aviary at New York Zoo in the early 1900s. It is no longer in use.

ABOVE The new Bird House at Frankfurt Zoo. Indoors the birds are seen against simulated natural backgrounds.

The 80-foot high Snowdon Aviary at London Zoo.

TOP OF PAGE The walls of the bear enclosures at St Louis Zoo are copies of a local rock formation.

ABOVE Man-made cavernous mountain for tigers and bears at Peking Zoo.

OPPOSITE The Antelope Yards at St Louis Zoo.

Wolves in Philadelphia Zoo's wolf woods.

Visitors at Washington Zoo find their way around with the help of these 'totem pole' signposts.

TOP OF PAGE Moats and well-hidden barriers are used to separate visitors and animals at East Berlin Zoo.

ABOVE Jurong Bird Park's free-flight aviary and man-made waterfall.

contribution, plus $7,000 for roads and $10,000 for an Elephant House – the building estimate for which was $20,000. The public still had reason to complain about the lack of a restaurant and of lavatories, and the keepers about their low pay.

Professor Langley died in 1906, and his successor as Secretary of the Smithsonian, a Dr Charles Walcott, who was a geologist and palaeontologist, neither shared his enthusiasm for the Zoo nor pleaded its cause with any conviction. In 1925 Dr William Mann was appointed Director. An entomologist by training, and an able administrator with a powerful personality, Dr Mann pushed through two phases of rebuilding, the first in 1927 and the second in 1935 – the latter in spite of the prevailing deep economic depression. A Bird House was built first, followed in 1931 by a Reptile House, and by a number of smaller works. In all, the first phase of building, which was completed just before the Depression began to take effect, cost about $370,000. Further money was authorized by Congress in 1935 for an Elephant House, a Small Mammal House, an extension to the Bird House, as well as a workshop and garage building. The cost of this phase of building turned out to be about $870,000, and through Dr Mann's skill in taking advantage of President Franklin Roosevelt's Public Works Programme, one of the measures he introduced to lift the USA out of the Depression, many other minor improvements were made possible. Even during the Depression the number of visitors never dropped below two million a year, and by 1938 it had risen to over three million – a tribute to the Zoo's public appeal. In 1944, by which time the USA was deeply immersed in the Second World War, numbers dropped to a low point of 1,803,532, and the annual budget to $277,130. The return of peace was associated with a phase of austerity during which no capital works could be started. Congress was, however, realistic enough to ensure that the yearly authorization for the maintenance of the Zoo kept in step with inflation. In 1949 it stood at $500,000.

In 1956, Dr Mann was succeeded as Director by Dr Theodore Reed, a veterinarian. A programme of redevelopment and re-organization began as soon as Congress provided adequate funds. Then, when the direct fiscal responsibility for the Zoo was transferred from Congress to the Smithsonian, Dr Reed reiterated that the purpose of the Zoo was 'education, conservation, and the advancement of science'. In 1975 the Smithsonian formally acquired some 3,150 acres (1,280 hectares) of pasture and woodland near Front Royal in Virginia as a Conservation and Research Centre, in keeping with the promise of the Zoo's 1971 *Annual Report* which had stated that 'Zoos are challenged to become net producers rather than net consumers of wildlife'.

Until 1941 the Zoo's staff had not included a veterinary surgeon. Only since 1956 have regular accounts of veterinary and pathological matters appeared in the Zoo's annual reports. A permanent Veterinary Department was established in 1963, but it had to work in temporary accommodation until 1970 when the Zoo was provided for the first time with an animal hospital and research building, into which the Zoo's Resident Scientist, Dr John Eisenberg, moved after having spent five years in two rooms on the upper floor of the Reptile House. Forty-four out of eighty-nine pages of the *Annual Report* for 1970 are devoted to an account of various enquiries, including a note about the occurrence of lead poisoning in monkeys and parrots due to paint that had been used on old cages. With increases in the number of its scientific staff, the Washington Zoo was at last in a state to fulfil one of the purposes laid on it in the 1889 Act. It had joined the small band of zoos where more than lip service is paid to the term 'scientific enquiry'.

The scientific staff are also closely involved in an active keeper-training programme which was introduced on a trial basis in 1969. Educational programmes for visitors and schoolchildren have also been properly organized, helped by a body called Friends of the National Zoo which was founded in 1958. Its members enjoy certain privileges, and are also trained to act as guides for parties of visitors, including students. In 1966 the Friends, of whom there are now more than 12,000, were authorized to raise funds by operating commercial concessions, and the organization is now responsible for catering, zoo shops, parking and a zoo train. As the five million people or so who now visit the Zoo annually do not pay an entrance fee, the funds which the Friends raise are an essential part of the Zoo's budget. Some is being ploughed back to pay for more formal educational courses which are held in the new Administration and Education Building. The Friends have become an established and important aspect of the National Zoo and of the Washington social scene.

Dr Langley gave the Park a good start in 1890 when he commissioned Washington's foremost landscape architects, Frederick Olmsted and his stepson John, to advise how the Zoo should be laid out. The 168-acre (70-hectare) site which it then comprised was only a small part of the steep-sided Rock Creek valley, through which a stream runs into the Potamac River. The Olmsteds designed a system of meandering roads and walkways that skilfully used the contours of the site, their design being of course planned for horse-and-carriage drives, for leisurely strolls and for a small number of visitors. None of the buildings that were put up over the first seventy years of the Zoo has any significant architectural merit. Most are pleasant brick-and-tile pastiches dressing up simple plans and economic constructions. The 1928 Bird House was designed by Howland Russell in a style reminiscent of Frank Lloyd Wright. After Russell's death the building was completed by Mr A. Harris, Architect for the District of Columbia. After his return from a study tour of zoos in Europe which he made with Dr Mann in 1929, Harris also designed a Byzantine-style Reptile House. Buildings put up in the late thirties were the work of Edwin Clarke, a specialist in zoo architecture. Soundly built, most of the pre-war houses survive, even though altered to meet the changing needs of the Zoo.

The first move towards a comprehensive redevelopment of the Zoo began in 1961 when the firm of architects and engineers, Daniel, Mann, Johnson and Mendenhall, were asked to prepare a master plan for the whole site. The Zoo's authorities, the Smithsonian Institution and the Commission of Fine Arts accepted their proposals in principle, and during the next few years the area between the Elephant House and Connecticut Avenue was rebuilt, with, in addition, an animal hospital and research building on the east side of Rock Creek. The Vietnam War then delayed further work but the Zoo nevertheless appointed its first staff architect to coordinate development in the hope that plans for the rebuilding of the central area of the whole site would be ready for implementation as soon as restrictions were lifted. The Commission of Fine Arts did not, however, favour the proposals that were then put forward, and the scheme was abandoned. The architects Faulkner, Fryer and Vanderpool were then appointed to review the situation and, mindful of the Commission's wish to 'emphasize the exhibition of animals in well-designed landscape settings and minimize the visual impact of the buildings', they brought in a landscape architect, Lester Collins, as consultant. At the same time engineers were commissioned to study parking and traffic problems. By 1972 a combined master plan had been prepared by all these consultants, in collaboration with Zoo officials. It received the approval of the Commission of Fine Arts and, in 1973, of the Smithsonian Institution.

The master plan is simple and clear, and expands on the Olmsteds' original concept of enhancing the natural character of the site. New buildings are to be subdued and screened by planting or by sinking them into the landscape. Usable old buildings will be retained and modernized to provide 'atmosphere'. At first sight the plan hardly seems to have changed the Park, yet it radically alters the roads and parking, so as to make the vast number of cars that now come to the Zoo less intrusive. It also subtly improves and modifies nearly every other aspect of the site. Now that the Friends are responsible for the sale of souvenirs and for catering it has been possible to extend this policy to areas where, in most zoos, design is conspicuously lacking. The plan will leave nothing to chance in providing an attractive environment.

The hope was that it should take no more than ten years to implement the new master plan, and there seems a good chance that this will be realized. Most of the work of construction and modernization has been carried out by, or in collaboration with, the original consultants. So far a new Lion and Tiger Complex, named after Dr Mann, has been built, and new kodiak and polar bear exhibits completed. The interior of the old Monkey House has been completely redesigned, and new outdoor enclosures constructed around the 1937 Elephant House for elephants, rhinos, giraffes and hippos. In 1977 the new Aministration and Education Building near Connecticut Avenue came into use, together with a fully equipped pathology unit. Under construction now are a new quarantine building, another major animal complex for the great apes, and a series of smaller exhibits for species like seal, otter, beaver and wolf. The new architecture of the Washington Zoo, paid for with federal money, is setting fresh standards for zoo building everywhere.

To sum up, the national zoo of the USA has had a strange history. For long periods it was very poor. Today it is rich. It began with great ideals, which were sometimes completely forgotten. It has been showered with gifts, and embarrassed by many. Its visitors do not pay to come in, but a 'Friends' organization is necessary to help out with funds. It is looked up to as a leader by other zoos, but has rarely been in the lead. There is a lot to learn, and enjoy, from this diversity of experience. But of one thing we can be quite certain – Washington has a truly great zoo.

NEW YORK

New York Zoological Park and Aquarium

Marcia Edwards

The origin of the New York Zoo, generally known as the Bronx Zoo, is closely associated with names that are celebrated in the history of the United States.

In 1887 Theodore Roosevelt, who fourteen years later was to become President of the United States, founded a club called the Boone and Crockett. Its members were men who, like himself, were passionately interested in big-game hunting and at the same time well aware of the damage that was being done to the native American fauna by uncontrolled and unprincipled hunting methods. As prominent citizens and men of considerable means, they tried to bring about the prohibition of the unsportsmanlike practices which they so deplored. In 1893 they were joined by a new member, Madison Grant, who had studied law at Columbia University and who, like Roosevelt, was involved in the reform of the administration of New York City. Although an energetic sportsman and in full accord with the principles of the Club, he had another idea in mind. He wanted a zoological park for New York. 'There was no other Zoology in the City worthy of the name and no opportunity to study Zoology, whereas in twenty-two cities of Europe and Asia and in two cities of America, the Zoological Garden formed a chief center of attraction, education and recreation.' If Washington and Philadelphia could boast a zoo, why not New York?

At Grant's suggestion, Roosevelt appointed a small committee of the Boone and Crockett Club to see what could be done. A charter for a Zoological Society was granted but development of the park was not easy. New York City agreed to provide the site, to prepare the ground and to maintain the Park when established. In return, the Society was to construct all necessary buildings and to purchase animals. But the City insisted that the Society should first raise $250,000 towards the cost. An Executive Committee was formed to campaign for funds, with Madison Grant, who had been Chairman of the Promotion Committee, as Secretary. He was fortunate in the support he received from Henry Fairfield Osborn. A professor at Columbia University and head of the Department of Vertebrate Paleontology at the American Museum of Natural History, Osborn was one of the leading scientists of his day and a man of some wealth and much influence. In the list of those who responded to the Society's appeal may be seen such familiar names as Andrew Carnegie, John D. Rockefeller, Cornelius and William Vanderbilt and William C. Whitney. Andrew H. Green, the so-called 'Father of Greater New York', became the Society's first President but he soon resigned, apparently because he felt that the proposed park was too grandiose. He was succeeded by Levi P. Morton, a former Governor of New York and one time Vice-President of the United States.

The objects of the Society were the establishment of a zoological park, the preservation of the native North American fauna and the promotion of zoology. The average European zoological garden comprised only about 30 acres (12 hectares), and limited space necessitated fairly close confinement of the animals. Grant and his colleagues had something very different in view. What they wanted was a place where 'the living creatures can be kept under conditions most closely approximating those with which nature usually surrounds them, in spaces so extensive that with many species the sense of confinement is either lost or greatly diminished, yet at the same time sufficiently limited that the animals are not inaccessible or invisible to the visitor'.

In April 1896 William T. Hornaday was appointed Director of the future park. He was well known as a naturalist and collector, and had been the first Director of the National Park in Washington. He at once began a survey of possible sites, and soon saw that an almost perfect location existed in South Bronx Park. The area, some 260 acres (105 hectares), included heavy forest and open woodland, hills and valleys, meadows, river and lake. Hornaday realized that the diversity of the terrain was well suited to the diversity of the animals it was intended to accommodate. Rocky outcrops could be used in animal enclosures, hollows turned into ponds, and the old-established trees would screen the necessary buildings.

The application for the land was successful, and the Society placed on record its opposition to the cutting-down of living trees in a public park whose natural aspect it was determined to preserve. The only departure from this sentiment was 'that the beavers shall be allowed to cut down several small trees that stand in the bog where it is proposed the Beaver Pond shall be located'. If this sounds like favouritism for beavers, it was countered by a later, somewhat stern instruction: 'The beavers will be required to build all the necessary dams for their ponds, all their canals, save one, and also their houses for winter use.' But the creatures were not always cooperative and it was 'on account of the annoyance and disappointment to visitors caused by the persistence of the beavers . . . in keeping out of sight' that the authorities conceded defeat and kept a pet beaver on show in the Small Mammal House.

In 1896 Hornaday went on a tour of zoological gardens in Europe, informing those he visited about the advantages of the site chosen for the New York Zoological Park and about the plans for its development. The report of his tour ends with the sentiments expressed by several zoo directors: 'With such ground, and the money that New York will give you, you can do anything that you choose.' They chose to do a number of things.

The first years of the Society's existence were spent constructing buildings and acquiring animals, in preparation for the opening of the Park towards the end of 1899. The larger exotic mammals were already well known to the public from travelling circuses, and so it was decided to concentrate initially on less familiar animals. Accordingly, the first building to be completed was the Reptile House for alligators, crocodiles, turtles, tortoises, lizards and snakes. It was judged a great success. The Curator, Raymond Ditmars, made many expeditions to secure animals and the New York reptile collection has always remained one of the largest of its kind in the world, not only in numbers but in the variety of forms which it includes.

The next two buildings, a great Flying Cage and an Aquatic Bird House, complemented each other. The former was spacious enough for flocks of various water birds, but because many of the tropical forms could not stand the New York winter, an indoor flight cage was built where they could be displayed during the cold months of the year.

A Primates House soon followed, then the Bear Dens in which young bears were reared, care being taken that the animals did not develop a bad temper. 'In the preservation of ursine temper', Hornaday reported, 'much has been gained by strictly prohibiting the feeding of bears by visitors.'

Next came the Lion House and quarters for other animals. In 1908, the Elephant House, at the time generally acknowledged to be the finest zoological building in the world, was opened. In addition to other unique features, each cage was supplied with currents of warm air for the greater comfort of the animals. The authorities were confident that the animals *enjoyed* their new abode, 'for were it otherwise', Hornaday wrote, 'the Indian rhinoceros would not, almost daily, gallop round and round, and with ponderous agility often leap into the air.'

Soon after the Zoo had started, it was suggested that the New York Aquarium, located in Battery Park, should be transferred to the management of the Zoological Society, to enable it to 'be placed on a permanent and scientific basis and be brought into closer touch with the educational system of the city'. The Society was reluctant to take on the responsibility but, in 1902, finally agreed. The Aquarium building had started out in 1807 as a fort. Then it became in turn an amusement centre, an immigrant landing station and, finally, a public aquarium. In taking it over, the Society stipulated that the city should provide adequate support for the institution and for the extension of its collections. Charles H. Townsend, who had worked for many years with the United States Fish Commission, was appointed the first Director of the new enterprise.

From the start the Zoo made a practice of acquiring animals that had not been seen before in America. Early on there was an extraordinary rarity from Ecuador, the first live spectacled bear to be shown in North America. Then there were three pygmy hippopotamuses. In the 1920s a duck-billed platypus, that strange Australian mammal which lays eggs yet suckles its young, was exhibited for the first time ever outside its native country, although unhappily it did not survive for long. Much later the Society kept a pair of these animals for over eleven years. Another 'first' for any zoo were the two giant monitor lizards, or Komodo dragons, which arrived in 1926. The Aquarium also claimed its share of unusual exhibits, including a manatee, one of the water-mammals generally believed to be the basis of the mermaid legend.

One of the original objects of the Society was the preservation of native North American animals. The indiscriminate slaughter of indigenous creatures had resulted in the annihilation of some species and had brought others to the edge of extinction. The buffalo herds were decimated and many birds fared as badly. It was evident, as the New York Zoological Society pointed out, that 'no species is too large, too small, too worthless or too remote to be sought out and destroyed by gun, trap, net or poison.'

The Society was energetic in its efforts to rouse public opinion in support of conservation and, in order to influence the legislature, established in the Zoo a department for game protection. A campaign to prevent the use of the automatic gun was partially successful. Measures were initiated to protect game in Alaska and British Columbia, and the attempt to create game sanctuaries was ceaselessly pursued. One notable victory, due to the combined exertions of the Society and the National Association of Audubon Societies, was the introduction of legislation to prohibit the importation, for millinery purposes, of the plumage and skins of wild birds.

In an effort to conserve the vanishing fauna, large areas of the Park were set aside in order to establish breeding groups of animals. The original herd of buffalo in the Zoo numbered twenty-five, and by 1907 had prospered sufficiently for fifteen animals to be sent to the Wichita Forest Reserve in Oklahoma, which the Government had designated a protected area. Over the years more bison were gathered and transported to other newly created reserves in Montana and South Dakota, thus ensuring the survival of the species.

Now, in common with many other zoos, New York has a programme to form self-replenishing breeding herds. Animals are loaned for breeding purposes and records are kept in order to maintain good blood lines. The Rare Animal Range and the Wildlife Survival Center are two recent ventures. The former, established in the Park in 1973, consists of large moated areas for herds of Père David's deer, wisent (or European bison), and Przewalski's horse. The Wildlife Survival Center, set upon St Catherine's Island off the coast of Georgia, is a cooperative venture of several American institutions, 'dedicated to the study of native wild animals through a long-term scientific research program, and to the propagation of all endangered species for which the island's climate and facilities may be suitable.'

But all flesh is as the grass, and the inmates of a zoological park are no exception. To improve facilities for the care of the animals and for the investigation of causes of death, a medical department was established in 1901, and an animal hospital was opened fifteen years later. Much attention is now given to preventive medicine. With the publication in 1964 of *The Management of Wild Animals in Captivity*, Lee S. Crandall, General Curator Emeritus of the Zoo, made available his knowledge gained from many years of practical experience. Today the Park has a large Animal Health Department and also seeks advice from specialist consultants.

The more specifically scientific work of the Society began with observations on wild sheep in Alaska. Later studies concerned the habits and anatomy of the animals in the Park, fish cultivation, and the diseases and parasites of fish in the Aquarium. In 1916 a Tropical Research Station was established at the suggestion of Dr William Beebe, Curator of Birds, in what was then British Guiana. The idea was that the station would work on the ecology of the animals in the area and would also act as a source of supply of animals for the Park. The scheme was a success and field work was extended to Bermuda and Venezuela. Much public interest was aroused by the first dives made in a bathysphere, a little vessel especially designed to observe life in the ocean depths. Then in 1949, as a permanent base for his field research, Beebe bought a property – 'Simla' – in the Arima Valley in Trinidad, which he later gave to the Society.

Henry Osborn's son, Fairfield, became President of the Society in 1940. He was much interested in many of the issues which now form the core of today's environmentalist movement, and in 1948, as a body parallel to the Society, established the Conservation Foundation, dedicating it to the 'protection and wise use of life-supporting resources'. A research programme into animal behaviour was initiated which then blossomed, as a joint venture with the Rockefeller University, into the Institute for Research in Animal Behaviour. This then incorporated the Zoo's Department of Tropical Research. 'Simla' was closed as a field station in 1971, and became part of the Asa Wright Nature Center, designed for the information and recreation of the people of Trinidad. In 1972, the animal behaviour research and the conservation interests of the Society became combined as the Center for Field Biology and Conservation. Under its leader, Dr George Schaller, this body continues a wide range of studies including the effects of pesticides, the management of wildlife in Africa, and the natural history of a variety of animals. Field research nearer home was not neglected, and in 1945 the Society organized a wildlife and conservation project, of research into the indigenous fauna, at Jackson Hole Wildlife Park in Wyoming. This station has since been transferred to the care of the University of Wyoming.

Under its first Director, Charles Townsend, the Aquarium had been vastly improved, but success brought difficulties. The building was antiquated and in bad condition and the aquarium space too small. The fire room, essential for heating water for tropical specimens, was liable to serious flooding during high tides. 'At such times', Townsend wrote, 'the firemen wear rubber hip boots and shovel wet coal into the furnaces from half-submerged wheelbarrows.' Over the years some improvements were made but a new building was still badly needed. In 1941 the construction of the Brooklyn–Battery Tunnel began, and as the Battery Park had to be closed to the public, the Aquarium had to move. Most of the collection was dispersed, but for many years a temporary aquarium was accommodated in the Lion House in the Bronx.

Eventually a 15-acre (6-hectare) site was made available on Coney Island, and the first stage of a new Aquarium was opened in 1957. The same year saw the start of a Department of Marine Biochemistry and Ecology for the study of marine environments, a subject relevant to the economic productivity of the sea. In 1965 this Department became part of the Osborn Laboratories of Marine Science, named in honour of Dr Fairfield Osborn, and comprising the Aquarium Laboratory, the Laboratory of Marine Biochemistry and Ecology and the Laboratory of Fish Genetics.

From the outset the Society concerned itself with the organization of lecture-demonstrations to school children, and New York's Board of Education was so impressed that it urgently, but fruitlessly, requested the city to provide the Park with a permanent lecture hall, so that children from the whole borough of Manhattan could participate. Now, many thousands of children come to the Park each year and the staff of the Education Department visit numerous schools to give lectures.

The Society realized that there was no guarantee that it would succeed in its campaign to protect the native fauna, and accordingly, in 1906, started a collection of notable specimens of heads, antlers and horns of large game animals, which in 1922 was housed in a new building, the National Collection of Heads and Horns Museum. Eventually the presentation of such trophies was deemed inconsistent with the Society's programmes and the Museum was closed in 1963. The Society also decided to commission oil paintings of endangered American animals. An impressive art gallery now contains many pictures by Carl Rungius, Charles Knight, Bruce Horsfall and Louis Agassiz Fuertes. There are also sculptures. One, a life-size bronze of the gorilla Dinah by Eli Harvey, commemorates a well-known inhabitant of the Park.

The publications of the Society started in 1896 with a news bulletin which contained details of the charter and by-laws, the application for the Bronx Park site and the names of officials and members. The bulletin became a regular feature through which members were kept informed of the progress of the Zoo. Its title was changed in 1942 to *Animal Kingdom*, which in recent years has been produced in conjunction with many other American zoos.

In 1897, the first *Annual Report* of the New York Zoological Society was published; until 1907, these reports contained accounts of the Society's scientific work. Then, a new journal, *Zoologica*, was launched which continued until 1971. To accommodate the results of the medical work at the Park and the Aquarium, another journal, *Zoopathologica*, was published between 1916 and 1928.

Notable among other publications is Beebe's *A Monograph of the Pheasants*. This work was the result of an expedition which Beebe made to the Far East to study pheasants in their natural surroundings. The book contains many beautiful illustrations and was issued in four volumes between 1918 and 1922.

From the start, the Society endeavoured to incorporate in its buildings the best in contemporary design. To maintain this ideal, a rebuilding programme was launched nearly fifty years after the Park opened, and many of the original houses have since been completely remodelled. An African Continental Enclosure, consisting of an open plains area, some 5 acres (2 hectares) in extent and separated from the public by moats, was officially opened in May 1941. In it are kept zebras, antelopes and African birds. A hidden moat separates it from Lion Island, on which lions appear to roam at liberty. In 1973 a similar display was created for such South American animals as tapirs, guanacos and rheas, and in 1977 a 40-acre (16-hectare) site was designed as a geographical exhibit of Asiatic mammals including deer, Indian rhinos and elephants, peacocks, cranes and Siberian tigers. The animals appear to be quite unrestricted, and visitors can view the whole scene from a 2-mile (3-kilometre) monorail excursion.

The Great Apes House, built in 1950, now provides quarters for gorillas. Moats confine the animals to their outside play areas, and inside the building glass partitions allow uninterrupted views of the inmates. Monkeys are now housed in the completely modernized old Primate House.

A new exhibit for penguins was opened in 1950. It is built on two levels, with a glass-sided pool which makes it possible to see the birds both on land and under water. The same device is exploited in the new Aquarium, where visitors can see both fishes and such marine mammals as walruses, whales and seals, in their underwater environment. In 1976 the Native Sealife Building was opened, planned especially to show the variety of the local fauna and flora.

It is only in recent years that zoos have exhibited nocturnal animals in conditions where they are awake and active. The technique is to reverse the normal lighting regime – by day the animals are kept under dim or red lights, by night under bright lights. New York's 'World of Darkness', opened in 1969, two years after London Zoo's 'Moonlight World', is designed to show nocturnal mammals, birds, reptiles and amphibians in such simulated habitat settings as an African forest floor, a swamp, a South American jungle and a tropical cave. The backgrounds for the animals in 'World of Darkness' were the work of the Zoo's Department of Exhibition, set up in 1963, and the first of its kind in an American zoo. It specializes in providing exhibitions, illustrative and educational material, to help the public to a better understanding of the animal collections.

There is also a new Aquatic Birds Building, dating from 1964, which rests on the foundations of the old Aquatic Bird House. Its basic purpose is to indicate the various types of birds found in wetland habitats and their dependence on such areas – ecology with a conservation message.

The 'World of Birds', opened in 1972, is a beautiful, imaginative exhibition created jointly by Mr William Conway, General Director of the Park, and a team of ornithologists, artists, designers, architects and gardeners. The cleverly planned two-level building again employs the habitat theme to show birds in their natural surroundings. Cliffs of artificial rock, 50 feet (15 metres) high, made from moulds of an actual cliff-face, were constructed for the New World rain forest. Cypress trees adorn the swamp, and jungle, scrub and tree-top are all represented to show birds from many continents and environments. Special exhibits explore habits, structure and distribution, and provide an insight into this diverse group of animals.

'Because of the undisputed fact that the Zoological Park is not merely a local institution, the Society has strongly objected to its being locally nicknamed the "Bronx Zoo", or "Zoo" in any form, with or without a prefix'. That was in 1901. Now, no one seems to mind, and the many changes over the years reflect the Zoo's determination to continue and to fulfil the purposes for which it was founded.

PRETORIA

National Zoological Gardens of South Africa

Colin Rawlins

A zoo could be regarded as superfluous in those fortunate parts of the world where wilderness areas are not remote from centres of human population. Such a country is South Africa – 'the greatest wildlife show on earth', as it has often been called. For hundreds of years there was wildlife to be seen everywhere, but by the second half of the nineteenth century it had become clear that the growing number of ever more efficiently armed hunters was proving destructive to so many species that legislation would be needed if extermination was not to become the rule. Accordingly, a law was passed in 1858 in the then South African Republic (later to become the Transvaal) to control the killing of game, including elephants. A further step was taken in 1898 by President Kruger when he proclaimed a vast area in northern Transvaal as the Sabie Game Reserve, today better known as the Kruger National Park. The movement to conserve wildlife can, therefore, be said to have started in South Africa, and the theme of conservation has never ceased to figure in South African national affairs. Today legislation to protect wildlife is comprehensive, and the Provincial Nature Conservation Ordinances of the Transvaal and Natal, where the main zoos are situated, provide not only for the protection of wild flora and fauna but also for the control of zoos and private game reserves.

The Pretoria Zoo dates officially from 1899 when a small collection of live animals belonging to the State Museum of the South African Republic was moved from the Museum premises to a nearby property called 'Rus in Urbe'. The Zoo came into existence mainly through the enthusiasm and determination of its first director, Dr P. W. B. Gunning.

Dr Gunning was also the Director of the State Museum, itself founded in 1892. It had been intended from the start that the Museum should include a living zoological collection, and animals were indeed being constantly brought there as gifts. Unfortunately the Jameson Raid at the end of 1895 temporarily prevented the development of Rus in Urbe as a zoo. The house was turned into a hostel for boys belonging to the local state school, and the land put under cultivation. Paradoxically it was the start of the Boer War in October 1899 which gave Dr Gunning the chance for which he had been waiting. The boys in the hostel moved out, either to join the forces or to go home, and he immediately got permission to move his small collection of animals from the yard behind the museum, where they had been kept, to Rus in Urbe. They consisted of a Serval cat, a yellow mongoose, a striped polecat, a common duiker, a Cape oryx or gemsbok, a Vervet monkey, a black-backed jackal, a Nile water monitor, a tortoise, an owl, two dormice, five baboons and about fifty small birds – less than a hundred animals in all. In this way the Pretoria Zoo began its long and distinguished career.

The occupation of the new site was intended to last only 'for the duration', but after the war the Committee of Management of the Museum confirmed the tenure of the site acquiring additional land across the small Apies River, which then formed the northern boundary of Rus in Urbe. In 1935 another piece of land, also north of the river, was added, making the total area of the Zoo today about 80 acres (32 hectares).

With the end of the Boer War, the Zoo developed rapidly under the energetic and devoted direction of Dr Gunning, who remained in charge until 1913. But even before the war the Zoo had made headline news. Dr Gunning had attended an agricultural congress in Cape Town and had met Cecil Rhodes, who had already set up his own small zoo near his house, Groote Schuur, in the Cape Peninsula. Rhodes generously offered a lioness for the Pretoria Zoo. With relations between Kruger in Pretoria and the British in the Cape deteriorating fast, Dr Gunning realized that it would be politically inexpedient to accept the gift. He therefore offered a silver pheasant and two eagles in exchange, and a deal was made. Gunning then set off by train for Pretoria with the lioness, but word had gone ahead that he was on his way with a British lion. He was greeted by a hostile reception at Pretoria Station, and received an official rebuke from the Secretary of State, who ordered him not to put the animal in the new Zoo but to keep it out of sight at his own house. The Museum authorities were even less tolerant, and demanded the creature's immediate return to Cape Town. The caged lioness was duly put back on a train which left the Republic just before the war started. The incident was officially recorded by the Republic's Director of Education, who was also Chairman of the Museum's Board, in the words: 'The notorious lion has been sent across the border together with other undesirables.' Rhodes then sent the animal to the London Zoo, from where she was returned to Pretoria after the war, arriving in July 1902, with her name changed from Fanny to Beauty. She lived until 1916 when she had to be destroyed because of blindness and other debilities of old age.

Dr Gunning not only increased the number of animals but also persuaded the Transvaal provincial government to provide funds for their maintenance. In 1913 the Union government decided that the Museum and Zoo, which until then had formed one institution for administrative purposes, should be separated. The Zoo then started on its independent existence as the Transvaal Zoological Gardens, with its own committee of management. Dr Gunning was duly appointed Director, but died within three months. In 1933 the Zoo became a state-aided institution, and a board of trustees was appointed to replace the former committee of management. Today the Zoo, which in 1916 was named the National Zoological Gardens of South Africa in recognition of its national status, is administered by the Department of Cultural Affairs, with the Minister of National Education having the political responsibility.

The years after the First World War proved to be a difficult period for the relatively young Zoo. The government of the day was not prepared to increase its annual grant to the Zoo, which had been fixed in 1909 at £6,000, a sum which was in fact reduced in the latter years of the war. It was reduced again during the Depression of the early thirties, when the Zoo's financial problems were made even more difficult by a severe decline in the number of its paying visitors.

Fortunately the Depression was to have at least one beneficial effect. It led to the government providing funds for capital works to relieve unemployment, some of which were allocated to the Zoo for the development of the new area north of the Apies River. Work started in 1932 and, although the unemployment relief scheme ended two years later, other funds were allocated for continuing the work, which went on until the so-called Northern Extension was finished in 1940. This governmental help for capital works was the second which the Zoo received. £10,000 had been authorized in 1910 by the then Colonial Secretary in response to an application from Dr Gunning for funds 'in order to enable the Committee to bring the institution up to a standard worthy of the capital of South Africa'.

It was not until the sixties that further large-scale development began again. By that time the finances of the National Zoological Gardens had been put on a much firmer basis through a more realistic annual budget for operational expenses from the National Government. In 1976 the governmental grant amounted to nearly £400,000, covering about three-fifths of the operating costs of the Zoo. The remainder came from the admission fees paid by visitors. The increase in visitors over the years, and particularly in recent times, has been significant. From about 100,000 visitors annually in the first years of the century, the figure rose to about 350,000 by 1950, and had doubled again by 1975.

The inspiration behind the major expansion of the Zoo came from the Zoo's third Director, Dr R. Bigalke, whose distinguished term of service lasted from 1927 to 1962. He had succeeded the second Director, Dr A. K. Haagner, at a most unpropitious time, but was to hand over to his own successor, Dr D. J. Brand, a smoothly running and financially viable institution, together with plans for extensive modernization. Dr Brand has had the pleasant task of implementing these plans, and of adding more of his own. The Pretoria Zoo is now up to date not only as a zoological exhibition but also as a centre of scientific research.

The main feature of the development of the Northern Extension was a run of enclosures, later called the Carl Jeppe Terraces, on the slopes of the 200-foot (61-metre) high ridge which marks the Zoo's northern boundary. By pre-Second World War standards the enclosures were very spacious, and lions and other big cats and wild sheep and goats were exhibited against a natural background of acacia trees. The service areas, including handling cages, were hidden from sight under the road from which the public looked at the animals. On the other side of the viewing area an extensive enclosure of several acres housed South African herd animals.

Dr Bigalke's plans for a building which was to contain the offices as well as a veterinary hospital and laboratory were realized in 1962 soon after his retirement, and the building was named after him. From then on development was to emphasize the special role of the Pretoria Zoo in supporting the national conservation policy. Three large enclosures, all several acres in extent, were built for South African antelopes and other hoofed animals, perhaps the most spectacular being one of 7 acres (2.8 hectares) for white rhinoceros. Here this species was to breed for the first time in captivity. A fine new Hippopotamus Enclosure was also laid out in which the animals could be seen in the most natural and attractive surroundings, including part of the old bed of the Apies River, once itself the natural home of wild hippos.

The most important development of recent years has been a new Aquarium and Reptile Park, which was opened by the South African Prime Minister in June 1974. With funds provided by the South African government, the Director, Dr Brand, and his Board of Trustees have built what must be one of the world's finest exhibits of its kind. The design of the Aquarium draws upon the experience of designers and zoologists all over the world. It consists of thirty-eight conventional tanks, two dioramas representing zoogeographical themes and one large circular tank. There are both freshwater and sea-water tanks. The service area is well laid out for efficient and easy operation, and water temperature and supply are automatically controlled. The reptiles and amphibians are kept either in a series of well-planted enclosures or in glass-fronted cages. There are also laboratories in which the Zoo's research staff are making a special study of the diseases of South African reptiles and of the biology of fish species which are difficult to keep in captivity.

Today the Zoo houses one of the biggest collections in the world, with over 800 mammals, 1,500 birds and 800 reptiles, and an extensive collection of over 5,000 fish. It ranks among the first ten major zoological collections not only in terms of numbers but also of species which are exhibited. In particular it owns what is probably the largest number of antelopes in any zoo.

More than twenty species of South African antelope have been bred at Pretoria, including impala, springbok, blesbok, lechwe, eland, the white-tailed gnu or black wildebeest, the grey duiker, and the Cape oryx or gemsbok. Many of these species are now far less plentiful than they once were, and breeding them in captivity not only helps to build up stock which could be returned to the wild but also increases our knowledge of their biology.

It was primarily with conservation in mind that the National Zoological Gardens acquired some 10,000 acres (4,047 hectares) of land near Lichtenburg in the western Transvaal for use as a game breeding centre. This was opened in November 1974. Herd animals, and particularly the South African antelopes, live almost in a state of nature in the very large area now available, and under conditions that can be called controlled rather than captive. Some remarkable results have already been achieved. For example, more than 300 springbok, 65 blesbok, 65 white-tailed and 20 brindled gnu, 18 gemsbok or oryx and 30 impala have been bred. Scimitar-horned oryx from the Sahara Desert and banteng, a species of wild cattle from Indonesia, both endangered in the wild, are also breeding successfully at this new centre.

In 1971 a smaller but corresponding project was started when a hundred acres (40.5 hectares) of natural bush not far from Pretoria were donated to the National Zoological Gardens for research into the feeding, breeding and behaviour of the cheetah. This animal has become rare in most parts of South Africa, and up to about 1970 had been bred in only a very few zoos. Some twenty animals were acquired for the project, and breeding started within two years. In the wild, males and females meet only for mating, and following this pattern the sexes are kept separately, with a single mature female in an enclosure into which one or more males are released at the correct time for mating. The numbers of animals bred at De Wildt, as Pretoria's Zoo's Cheetah Research and Breeding Station is called, has increased each year. Thirty-four cheetahs were bred in 1976 and nineteen in 1977.

Pretoria Zoo has, of course, bred other interesting Southern African species. The little known aardwolf, something of a mixture between a jackal and a hyena in appearance, had rarely been seen in zoos and had never been bred in captivity before 1965, when a litter was born at Pretoria. The aardwolf feeds mainly on insects, particularly termites, and lives a nocturnal life, lying up during the day in burrows, sometimes in those made by its ant-eating colleague the aardvark, or 'earth-pig', so called because of the long pig-like snout with which it probes anthills. The aardvark has also been bred at Pretoria Zoo.

The Cape hunting dog (sometimes called the African wild dog), the black-backed jackal, the Caracal lynx and the Cape fur seal are some of the other interesting species bred at Pretoria. The breeding, many years ago, of the yellow-billed hornbill in artificially made tree-trunk nests in the Zoo's Large Flight Aviary was one of the earliest occasions when this still relatively rare technique was used.

The most celebrated achievement of the Pretoria Zoo was undoubtedly the first captive birth of a white rhinoceros. The term 'white' is an English corruption of the Afrikaans word '*wijd*', meaning wide or broad and referring to the shape of the animal's lips (it is also called the square-lipped rhinoceros to differentiate it from the black rhino which has pointed and prehensile lips). The white rhino had suffered more than any other large mammal of South Africa from the depredations of hunters during the 1800s, and by the early part of this century its numbers in South Africa had fallen to below fifty. The situation was then gradually improved, mainly because of the efforts of a few dedicated conservationists, particularly in Zululand. In 1946 a young female white rhino named Zuluana was sent to the Pretoria Zoo, the first time the species had ever been kept in a Zoo. Two years later a young male was acquired and then a second female. They became a popular and important exhibit, but did not breed. Zuluana, at the age of thirty-four, is now the oldest captive white rhino.

The failure of this first group to breed stimulated the Zoo authorities to try again, and a new group of one male and two female adult white rhinos was acquired in April 1966 and put into a new 7-acre (2.8 hectare) enclosure. In June of the following year a calf was born, the first ever in captivity, but since gestation for this species is believed to last about seventeen months, it seemed that conception had taken place in the wild. A second calf, clearly conceived in the Zoo, was born in 1969, and others have followed at regular yearly intervals. The success at Pretoria, followed by births in other zoos, began to show clearly that space, rather than the number of animals in the group, as had once been thought, was the decisive factor in white rhino breeding.

The support given by the State to the National Zoological Gardens, which has been particularly generous in recent years, has been largely due to its concern with education and research. The Zoo's large Education Department is staffed by professionally qualified teachers whose work is coordinated with that of the education departments, both European and African, of the national and provincial governments. The emphasis in the Zoo's educational work is on conservation of South

African flora and fauna. The Department has a first-class art section which is responsible for most of the excellent displays of information on animals and conservation. Pretoria Zoo is undoubtedly outstanding in the presentation of educational material.

The scientific research of the National Zoological Gardens is also held in respect by the scientific community of South Africa. A close working relationship exists with the provincial departments of nature conservation, with the universities, and with the renowned South African Veterinary Research Centre at Onderstepoort. The modern animal hospital at the Zoo has led to useful work on many aspects of veterinary science, and the study of the use of narcotics and sedatives for the handling of large wild animals has been a special aspect of this work.

South Africa has every reason to be proud of the way its National Zoo has developed and of the part it is playing, not only in the conservation of the country's natural fauna, but also in the growth of scientific zoology.

PEKING

The Peking Zoo

Michael Brambell

A visit to the Peking Zoo, famous for its giant pandas, quickly makes a European realize that China is at 'the other end' of the world. Yet while the Zoo's oriental aspect makes a major impression at the start, it is also clear from many of the exhibits that the authorities are concerned to display animals from all parts of the world. Before 1949 there were only three or four zoos in China, and all were badly run down as a result of the political unrest of the previous thirty years. The Peking Zoo then housed only ten monkeys, three decrepit parrots and a blind emu. The Shanghai Zoo, which had served in great part as a staging post for animals intended for export, was also in very poor shape. There had been a zoo at Chungking, and there may have been one or two in other parts of the country. The fact was that few of China's population had any idea of the wealth of wildlife within their own country, let alone the enormous wealth that existed elsewhere in the world.

Yet legend has it that there was a zoo in the Chinese capital as far back as three thousand years ago. Peking itself did not become the capital of China until Kublai Khan became emperor in 1279. It was during his reign that Marco Polo visited China and that trade links with the west of Asia and, either directly or indirectly, with Africa were opened up. A record exists that not much more than a century later giraffes could be seen in Peking, and it is claimed that the first zoo ever to breed giraffes was Peking, in the latter part of the eighteenth century. But while the Chinese were curious about animals from overseas, they did not seem to have been able to distinguish the polar bear from the giant panda, calling both 'bei xiung' – white bear. It was not until Père Armand David began his zoological and botanical exploration of China that people inside and outside the country awakened to the richness of its fauna and flora, which Arthur Sowerby described so well in the early part of our own century. Today there is an awareness, at least among Chinese scientists, that the animals and plants of that country are uniquely interesting, and that the continued existence of many species is dependent on strict measures of conservation.

Peking's present Zoo was started in 1906 principally for the enjoyment of dowager Empress Tse Hui. It was built in the Sanbeizi Park where imperial eunuchs had lived in the past. Later it became known, perhaps rather optimistically, as the Park of Ten Thousand Animals. In 1908 tickets were issued to visitors, though it was not clear how far down the social scale these tickets were available. At that time there were elephant, lion, ostrich, hyaena, puma and some zebra, brought back from Germany as particular curiosities by the Manchu minister, Tuan Fang.

When the Ching dynasty was overthrown in 1911 the Zoo was

converted into an experimental farm where, amongst other enterprises, silk worms and cotton were cultivated. Later, as the Japanese encroachment became more menacing, Chiang Kai-shek used the farm as a military camp, as did the Japanese when they occupied Peking. It was not until after 1949, when the Chinese Communist armies entered the city, that the Zoo was reinstated, and then opened to the public. In 1952 a new Elephant House was built on the site of the cotton fields of the experimental farm, and three years later the Zoo was officially named the Peking Zoological Garden, the name of the park in which it was sited having been changed from Sanbeizi to West Suburbs Park.

Since the proclamation of the People's Republic, major zoos have also been set up in some forty other Chinese cities, as well as many smaller zoos in other parts of the country. With five to eight million people entering the Peking Zoo every year, and no doubt staggeringly large numbers of visitors going to other zoos, it is clear that a sizeable proportion of China's immense population now has the chance to see indigenous and exotic animals.

The West Suburbs Park is bounded to the north by a river, from which water has been directed into a string of channels, ponds and lakes which run through the middle of the Zoo. The main gate leads into a courtyard where the administrative offices and other service buildings are situated. On a doorpost which faces the gatekeeper is painted a mark one metre clear of the ground. Any child who walks past the mark without obscuring the gatekeeper's view of it is allowed in free.

To the right of the main gate are small cages laid out ornamentally. They house snub-nosed monkeys, rabbits, foxes and a variety of other small animals.

Peking's famed collection of giant pandas is also kept near the main gate. Their house is very simple, with an indoor hall through which visitors can walk past barred cages that open on to a large outdoor walled paddock. There is a service area in the building, though access to the paddock for keepers is, if not over the wall, through the same gate as that used by the animals. The paddock is furnished with rocks and a climbing frame, as well as a swing.

The giant panda is found in the wild only in Szechwan, in western China, in the inaccessible and inhospitable steep-sided mountain valleys characteristic of the eastern edge of the Tibet–Tsinghai plateau. Giant pandas are shy and retiring animals, resting in the higher regions during the daytime and at night descending to the lower bamboo forests to feed. Szechwan is comparatively remote from the more populated parts of China, but it is a very fertile area and has been a part of historical China for a number of centuries. It is therefore somewhat surprising that there is no convincing record of any Chinese, except those who live in Szechwan, knowing of the giant panda until Père Armand David, the famous French traveller, was shown a dressed skin during the course of a trip he made to that part of the country in 1869. It was more than forty years after Père David's discovery that the first sighting of a live giant panda was recorded by the Stoetzmer Expedition. Then, in the 1920s and early 1930s, there was a veritable scramble for dead specimens. The first living giant panda to leave the country was a baby called Su-Lin which was taken to Brookfield Zoo in Chicago in 1936. Two years later five live pandas were exported from China, including Ming who went to London where her fame was so great that, when she died in 1944, *The Times* published an obituary.

After the Communist victory in 1949, the Chinese began to concern themselves with the preservation and scientific study of the panda. In 1955 the first three pandas arrived in Peking Zoo, and these were added to in the following years. Between 1957 and 1959 three pandas were exported, two males to Russia and a third, a female, Chi Chi, which ended up in the London Zoo. The first male which went to Russia died soon afterwards, but the second, An An, and the London female, Chi Chi, both survived to old age. After much diplomatic negotiation and some widespread and good-humoured publicity, the two were brought together first in Moscow, and then in London. But mating did not take place. Apart from a pair which went to North Korea, no more giant pandas left China until 1972, when pairs were sent as State gifts first to the United States and then to Japan, France, the United Kingdom, Mexico and Spain.

The two Peking pandas, Pi Pi and Li Li, were successfully mated, and in 1963 Li Li gave birth to the world's first giant panda born in captivity – Ming Ming. The panda is minute at birth, weighing only about 3½ ounces (100 grams), which is less than the weight of two hens' eggs. It is white, almost hairless, and has its eyes shut. The characteristic black and white coat appears about one month after birth and the eyes open two to four weeks later. The young panda continues to suckle for six months, and then becomes independent of its mother fairly quickly. Since Ming Ming's birth several more giant pandas have been born and reared in Peking and Shanghai Zoos.

Close to the Peking Panda House is a large lake edged with weeping willows. Australian black swans were bred here in 1963, seven years after the parent pair had arrived at the Zoo. One of the six islands in the lake is the home of one of the world's most comprehensive collections of cranes, with at least eleven of the world's fifteen known species.

Nearby are several paddocks for ungulates where, apart from a very good collection of exotic species, enough native species are exhibited to make a zoologist's mouth water. There are kiang, the east Asian equid, which are more closely related to the wild ass than to the horse. There is a large and thriving herd of white-lipped deer, a tall and elegant animal found only in China. The visitor would be lucky to see the stags in full antler for unfortunately no sooner is the velvet shed than the antlers are sawn off and sold to be made into traditional Chinese medicine. It seems that the only deer which does not suffer this fate is the fallow deer whose antlers do not provide any constituent of traditional Chinese medicine. There are also herds of river deer, musk deer and muntjacs.

The Carnivore Enclosure is the Zoo's most striking 'building'. It consists of several large concrete-surfaced paddocks surrounded by a high wall topped by a public walkway. When not in the open the animals, among them the South China tiger, the Chinese leopard and the Heilungkiang (or Amur) tiger, are housed in dens within a central concrete mountain. The Heilungkiang tiger is a magnificent creature which is thought to belong to the same sub-species as the Siberian tiger. In nearby enclosures are the bears, including a form of brown bear from Sinkiang which the Chinese call 'horse bear' and which again is not to be seen in any zoo outside China. The 'horse bear' is midway in size between the relatively small Caucasus brown bear and the much larger Kamchatka brown bear.

Not far from the carnivore enclosure is a new building for monkeys, which houses Peking's primate specialities. Among the rare animals are breeding groups of the snub-nosed monkey (*Rhinopithecus roxellanae*), also called the retroussé-nosed langur or the golden langur (not to be confused with the Indian species *Presbytis geei* which has been given the same common name). The snub-nosed monkeys live in the western foothills of the Tibet–Tsinghai plateau and are included in the primate sub-family in which are classified the langurs of Asia and the colobuses of Africa. All these creatures have evolved a multi-chambered stomach and digest green vegetation much as do cattle and sheep. What makes the snub-nosed monkey particularly interesting is that, unlike other monkeys and apes, most of which have tropical or sub-tropical habitats, it can survive the intense cold of the long winter of the mountainous area while living on twigs and bark. The only monkeys which live in a more northerly latitude are the Japanese macaques and the Chinese Tcheli monkey, both of which have simple stomachs and both of which live in much milder climates, even though they are further from the equator.

In addition to the snub-nosed Monkey, Peking Zoo displays a group of white-headed langurs (*Trachypithecus leucocephalus*) from the south of China. These animals have never been seen outside China and were, in fact, only discovered in recent years.

Another interesting Peking exhibit is a group of Père David's deer. Though once abundant and widely distributed in north-eastern China, the species ceased to exist in the wild at the time of the Shang dynasty (1766–1122 BC) when the swamps in which it lived were brought under cultivation. The only surviving herd was then preserved in the Imperial Hunting Park near Peking. Entry to the Park was strictly forbidden to foreigners, but in 1866 Père David bought from the park keepers two skins which he sent to Paris, where Milne Edwards allocated the creature to a species new to science. In 1894 the wall of the Park was breached by flood waters of the Hun Ho River, and all the deer escaped into the surrounding countryside, where most were killed and eaten. The few survivors that lived through the Boxer Rebellion of 1899–1901 were taken to the Park in Peking, but by about 1920 the species had died out in China. Fortunately, however, the eleventh Duke of Bedford had between 1900 and 1901 collected together in his park at Woburn in England all the Père David deer that had been shipped to European zoos at various times in the latter part of the nineteenth century. The Woburn herd soon grew and surplus animals were then sent to other collections including, in 1944, that belonging to the Zoological Society of London at Whipsnade. In 1956 four animals from the Whipsnade herd were taken to Peking. Their arrival was regarded as a matter of such importance that Thomas Sangster, a keeper from the London Zoo who had been in charge of the animals on their flight, was entertained by Chairman Mao Tse-tung and Prime Minister Chou En-lai, an honour that has been paid to no other zoo representative from outside China. In 1973 six more Père David deer were taken to China from Whipsnade and the two groups combined into one breeding herd.

The Peking Zoo also contains examples of the takin, a musk-ox-like animal found only in the Himalayas and China which, because of the inaccessibility of its habitat, is little known.

All species of the eared pheasants, the brown, blue and white, abound in Peking Zoo. The rarest, the white eared pheasant, may well have been saved from extinction by the action of the Zoo staff, when a clutch of eggs was taken from the wild and hatched under a domestic bird. Two pairs of the descendants of the successful hatch were sent to Jersey Zoo, and a thriving breeding group was established, so much so that several hundred of these birds now live in collections throughout the world.

China is the home of many species which are not found anywhere else. There is the Yangtse alligator, which can be seen in Peking and which, with the Mississippi alligator, is the only surviving species of its kind. In Asia its habitat is the riverline marshland along the lower reaches of the Yangtse. At the end of the summer when the river is at its lowest the alligators burrow into the drying mud to a depth of about 4½ feet (1.5 metres) and remain buried until the floods re-inundate the marshland.

An Aquatic Animals House has recently been built in the Peking Zoo, and among its first inmates was a pair of manatees from Mexico. They are fed water hyancinth which, in some areas of China, is so widespread that it has become a pest. The house will probably be used to display some of China's indigenous aquatic animals, among them the Chinese sturgeon, and the Chinese river dolphin. The latter is fast disappearing from the rivers of China and it is hoped that it will breed in the Zoo so as to ensure its survival.

The oddest collection of all in the Peking Zoo is the goldfish collection. There are rows and rows of half-barrels, each containing a different variety of goldfish. Very few of them are gold. There are black, red, silver and blue forms. Some have long tails, some have long fins, others have great pouches under their chins, and some have huge protuberances on either side of the head. In a fanciful way they are attractive.

The Peking Zoo is a municipal zoo and is the responsibility of the Peking Office of Parks and Forestry, a department of the Peking Municipal Revolutionary Committee. The Zoo is directed by a chairman and four vice-chairmen (the equivalent of deputy directors), each with a special responsibility. There is one head keeper, about seventeen zoologically trained technicians, and about 120 keepers who have learnt some zoology at school. There is also a veterinary department. The staff of the Peking Zoo are hard-working and dedicated. Because they have so little technical literature to consult they have little opportunity of finding out what is happening in zoos outside China, and even in the other zoos of China. The Zoo cooperates closely with the Institute of Zoology of the Chinese Academy of Sciences. There is little doubt that in the future Chinese zoos will play an even more important part than they do now in the conservation of the wildlife of China, both through breeding and through education.

ST LOUIS

St Louis Zoological Park

Michael Brambell

The St Louis Zoo is not the oldest in the United States, nor the biggest, nor the most diverse. It is, however, typical of most American civic zoos, being financed and run by the local authority, yet strongly supported by a Zoological Society that is much respected within the community. It is a good zoo with a spacious charm that is so often lacking in many newer establishments.

The Zoo started almost by accident. In 1803 the territories along the Mississippi River, then known collectively as Louisiana, and including the city of St Louis, were purchased by the United States from France. The centenary of the acquisition was marked by a World's Fair – the Louisiana Purchase Exposition – which took place in Forest Park within the city, and where the Smithsonian Institution of Washington had built a large aviary to mark its respect. It was only some ten years since Congress had decided that Washington should be graced with a national zoo.

In 1910 a Zoological Society of St Louis was founded and began to collect the money to buy the aviary for the city. Three years later the local authority allocated 77 acres (31 hectares) of the Park for a new zoo, and then provided funds to build a Lion House. A year later, after a local referendum, the Zoo was taken over by the city from the Zoological Society, which had been incorporated as an independent civic organization. An annual levy on the value of the property was then agreed, thus making St Louis the first city in the world to support a zoo by taxes imposed for that specific purpose. So it was that the Zoo became a civic amenity, administered by a Zoological Board of Control consisting of nine men. The Bill passed by the General Assembly of the State of Missouri in 1915 stated, *inter alia*, that 'every zoological park established under this act . . . shall be forever free to the use of the inhabitants of the city . . . so long as the zoological park tax shall be operated'. This arrangement provided the Zoo with a guaranteed source of income and ensured that it would always be open to the public.

But these provisions also had their disadvantages. As in many American zoos to which the public had free access, steps had to be taken by the police to keep out hooligans (motor-cycle police patrols are not an unusual sight in the Bronx Zoo). More important, the Zoo's income did not increase *pari passu* with its attractiveness. A similar financial problem in other American zoos has been solved by charging entry fees for specific exhibits, but this was not allowed at St Louis. Worse still, the value of the property on which the zoo tax was based was falling, due to the same process of urban decay which has affected so many cities – not only American – since the Second World War, as people and business have moved to the suburbs. The people of St Louis who

RIGHT Newly born Nyala antelope at Prague Zoo.

BELOW Bald eaglet hatched at Washington Zoo in 1978.

Koala bears climb on branches strewn with the day's supply of eucalyptus leaves at Taronga Zoo.

Giant pandas play at Peking Zoo.

ABOVE At San Diego Wild Animal Park waterfowl indigenous to East Africa can be seen in Nairobi Village.

OPPOSITE Barasingha deer in New York Zoo's 40-acre enclosure for Asiatic animals.

ABOVE The Indo-Malayan Pavilion housing delicate animals at Toronto Zoo.

OPPOSITE Polar bears in the North American Region at Toronto Zoo.

Visitors can see the animals – here, Indian rhinoceroses – in New York Zoo's Asiatic enclosure from the monorail.

had served the city's Zoo and museums so well were no longer there. But the community spirit certainly was. In 1971 another referendum was held to determine whether there would be agreement to the formation of a wider 'district' which would look after the city's Zoo and museums, and a Zoological Park and Museum District of the City and County of St Louis was founded, with a Zoological sub-district. With its funds now independent of the vagaries of politics, the Zoo is today on a sound financial footing. Other American municipal zoos have fluctuated, and continue to fluctuate, between expansion and penury with their cities' fortunes. Fortunately, the Zoological Society of St Louis continued to function as a voluntary body, supporting the work of the Zoo throughout the period of uncertainty and raising money for special projects. In 1977 the Society's name was changed to the St Louis Zoo Friends Association.

The Zoo occupies a block of land near the centre of the city, close to the Art Museum. The site slopes gently and is effectively divided up into four areas by an x-shaped series of lakes and ponds. The oldest part of the Zoo is towards the east, in which are found the original Great Aviary, a Bird House, an Ape House, and the Antelope House with its enclosures. Along much of the boundary of this part of the Zoo is a curving Jeffersonian brick wall, very much part of the American tradition since its invention by Thomas Jefferson, the third President of the United States.

On the north side of the Zoo are the Eagles' Aviary and the Bear Enclosures, walled by expensive castings of the Missouri Palisades, the rock formation on the banks of the great river not far from the city. In the southern part is what is called the New Big Cat Country, which replaces a conventionally built house that had been put up for lions, tigers and other members of the cat family. Nearby is the Primate House, which has also been completely renovated in recent years, and the Reptile House, to which is attached a wing for small mammals.

The newest and largest sector of the Zoo is to the west. Here are the Elephant House, a house for aquatic animals, an extensive Children's Zoo with, close by, the non-public service areas and an animal hospital. Around the Zoo runs a narrow-gauge railway with four stations on a 1.2-mile (2-kilometre) long track. The track passes through two long tunnels under the busy eastern and southern parts of the Zoo.

Birds are a special feature of the St Louis Zoo, which now keeps representatives of over 300 species belonging to 60 families. The Aviary which was put up by the Smithsonian Institution is still one of the largest aviaries in the world. In 1967 it was completely renovated and provided with better landscaping, a raised walkway, and two small lakes for wading and swimming birds. Among its 125 inmates, belonging to some 40 species, are flamingos, storks, cormorants, gulls, cranes, guinea-fowl, spoonbills, egrets, pheasants, ducks, geese, screamers, ibises, pelicans, penguins and a road-runner.

Even more impressive are the waterfowl lakes. These were restocked in 1964, and there are now about 450 birds of 90 species. An incubation unit was set up which handles approximately 1,000 eggs a year. With a fertility rate of about 66 per cent and a hatching success of about half of this, it is no wonder that St Louis is proud of its waterfowl.

St Louis also maintains a good penguin collection inside the Aquatic Animal House, with underwater viewing, cooling facilities, and air and water filtration. Most of the other birds the Zoo keeps are in the Tropical Bird House, which was built in 1930. It is broken up into several free-flying areas where the birds tend to stay in the protection of the foliage and away from the darker areas where the public walk. An especially good collection of South American birds is displayed in this house.

A representative collection of snakes and lizards, as well as of amphibians and fish, is kept in the remodelled Herpetarium, where a male tuatara lived from 1966 to 1972, so establishing an impressive record for this very rare New Zealand lizard. In 1973 the New Zealand government sent a replacement pair of tuatara after the Zoo had arranged an artificial environment which reflected as closely as possible the natural conditions in which the beast is found. The female soon began digging into the loose substrata of the cage and laid eleven eggs – none of which, unfortunately, hatched. Today there are only nineteen tuataras in eleven collections outside its native habitat. The St Louis Zoo authorities are skilled at breeding snakes and lizards and have, for example, developed a useful method of incubating reptile eggs by placing then on damp peat within plastic bags. Warmth is provided by the slowly decaying peat, and moisture is retained within the plastic. Once the eggs hatch and the young become active, it is necessary to open the bags to let in more oxygen.

But however impressive the St Louis birds and reptiles, the memory that the ordinary visitor will take away from the Zoo will be of the enclosures for mammals. Those for the bears must be among the best designed of their kind in the world, even if they are built of concrete made to simulate rock. Artificial rock also separates the paddocks of the Antelope House from each other. Here the rock faithfully reproduces the outcrops of wind-weathered blocks of granite seen at Graniteville, a town to the south-west of St Louis. Not only has the rock been copied, but care has been taken to plant the cracks between the artificial rocks with the same species of plants as occur at Graniteville. The art of making artificial rock is carried to perfection in several parts of America, perhaps most strikingly in the Arizona-Sonora Desert Museum at Tucson, Arizona. But it is a very costly art!

The twenty-five outdoor enclosures of the Antelope House contain, among other creatures, klipspringer, springbok, kudu, gaur, banteng, gerenuk, giraffe, Grevy's zebras and Speke's gazelle from Africa, and llamas from South America. The herd of the rare Speke's gazelle is especially noteworthy, containing as it did in 1977 sixteen individuals, of which fourteen were born in captivity. This interesting species, found only in the deserts of Somalia, has a wrinkled patch of skin at the end of the muzzle above the nostril. In both sexes this can be suddenly inflated into a balloon and as quickly deflated again – presumably to startle other animals.

The New Big Cat Country, which consists of seven landscaped and planted enclosures, was constructed to replace the conventional Big Cat House put up in 1917, and has earned the Zoo its second award from the American Association of Zoological Parks and Aquariums (AAZPA), the first having been given in recognition of the skilful rebuilding of the old Primate House, the interior of which was stripped so that thirty-three old barred cages could be replaced by thirteen glass-fronted large enclosures. The latter have been carefully landscaped to suit the species they house, and each is individually air-conditioned. A separate air-conditioning system also serves the public area. Since very few monkeys can tolerate the prolonged and severe winters which St Louis often experiences there are no outside enclosures.

In the south-west part of the Zoo is the house for elephants, rhinoceroses and hippopotamuses. The latter recently made zoo history by producing heathly twins.

St Louis is one of the few zoos described in this book which regularly puts on public shows of circus-trained animals. There is a new show every year by highly trained chimpanzees, and there are also performances by trained elephants and sealions. This combination of circus and zoo might affront people who are opposed to the training of exotic animals. But no one could do anything but praise the cooperation which the Zoo's officials have established between St Louis Zoo and the College of Veterinary Medicine, University of Missouri, Columbia. The Zoo is also used as the venue for courses in animal behaviour by the Universities of St Louis and Washington. St Louis was the first zoo in the world to set up a 'tissue bank', in which tissue removed from exotic animals dying in the zoo are taken and stored for study by scientists anywhere in the world. No doubt St Louis will help teach other zoos how to reconcile what the public likes to see with what is really the scientific justification for zoos in general.

SAN DIEGO

The San Diego Zoo
and
Wild Animal Park

Marcia Edwards

The sun shines on San Diego some 330 days of every year, an important factor in a story which started in 1916, the second year of the Panama–California International Exposition. But the two-year fair, which was held in Balboa Park, proved to be over-ambitious for a city of modest resources, and by the end of the second year many of the exhibitors had moved out. Some left behind small collections of animals which had been among the attractions of the show.

Harry Wegeforth, a well-known doctor with a practice in San Diego, was fascinated by animals and had long thought that the city needed a zoo. Prompted by the plight of the deserted animals, he persuaded *The San Diego Union* to publish an article outlining a proposal to promote a zoological society and garden. The response to his suggestion resulted in the formation of a board of directors and the subsequent foundation of the Zoological Society of San Diego. The state of the exposition animals had caused much public disquiet and the Board of Park Commissioners was only too ready to permit the new Society to take charge of the abandoned collections.

A permanent site for the new zoo was the next requirement, but that took time. Meanwhile the zoo needed accommodation for its staff, and Dr Wegeforth managed to obtain, as a gift, one of the disused exposition buildings. But then the Society was advised that it was not legally entitled to own buildings and it was therefore decided to give the building it had secured and any future buildings to the Park Department for the use of the Zoological Society.

Then the Park Commissioners discovered that a vote of the people was needed before public land could be given away. But by this time the United States had become directly involved in the First World War, and the question of a location for the Zoo was put aside.

It was not until the autumn of 1921 that over 100 acres (40.5 hectares) in Balboa Park were eventually leased to the Zoo, which by that time had transferred the ownership of its animals to the city of San Diego. The Zoological Society was thus able to administer the Zoo without owning the land, the buildings or the animals.

The area allocated to the Society consisted of high tableland – mesa – divided by steep-sided canyons, totally unsuited for normal park development and therefore considered of little value. But if useless for formal flowers beds, the site was admirable for the sort of zoo Harry Wegeforth had in mind. Plans for development were approved, and within a year the Zoo was properly installed.

The nature of the terrain enabled many of the animals to be displayed in enclosures with moats rather than bars to confine them. The plan was to set aside the mesas for hoofed animals, monkeys and birds, and the canyons for bears and large cats. The latter were to be housed in open, spacious, cave-like grottoes, where they could be seen without unsightly caging.

Miss Ellen Browning Scripps, a patron of many Californian institutions, including the world-famous Scripps Institute of Oceanography at La Jolla, was a great benefactress of the Zoo. She provided the funds not only to build the grottoes, but also to fence the Zoo grounds. This was a very practical gift, for without fencing the Zoo would not have been able to levy any admission charges. Another of her early gifts was a Flight Aviary for wading birds. Built across a canyon, it was completed in 1923, and at the time was much publicized as the world's largest aviary. Captain G. Allan Hancock was another generous benefactor, and his yacht *Velero III* was often put at the disposal of Dr Wegeforth and his colleagues for animal-collecting trips.

But if many public-spirited people helped the Zoo in its early days, so too did the climate. Dr Wegeforth and his colleagues were spared the usual formidable cost that has to be borne in most North American and European zoos where animals have to be sheltered in warmed winter quarters for several months of the year. The Californian sun and the mild winter not only made this unnecessary but also made it possible to display the animals in the open for most of the year.

Not surprisingly for a physician, animal health was one of Dr Wegeforth's main concerns and he wanted a hospital for the study and treatment of animal diseases. His wish was realized in 1927 when, as a result of the generosity of Ellen Scripps, a hospital, named in her honour, was opened.

When the Zoo decided to buy more animals, Harry Wegeforth discovered that the market was so managed that zoos were obliged to pay whatever the dealers demanded. Typically he decided to circumvent them, and it was not long before he had planned an organization through which zoo directors could keep each other informed about the animals they wanted and about those of which they had a surplus. So it was that at a meeting in California in 1924, a National Association of Zoo Executives was formed. Directors of aquaria soon wished to participate, so the title was changed to the American Association of Zoological Parks and Aquariums (AAZPA). The Association grew and it now has an international reputation.

Early efforts to become self-sufficient in animal supplies proved so successful that by 1924 the Zoo was able to make its first large shipment of animals. This went to Taronga Zoo in Sydney, so establishing a valuable link between the two institutions. In return came a number of Australian animals, including two koalas. This was in the days before the Australian government had imposed a ban on the export of most of its native animals, and although one of the pair soon died, the other lived to become one of the Zoo's most popular animals. When it eventually died, the Zoo tried to persuade the Australian authorities to send another pair, pointing out that the climate of San Diego was suitable for koalas and that the creatures' natural food would be provided by the many varieties of eucalyptus that had been planted in the Zoo grounds. The request was turned down, but many years later, in 1953, the Zoo's wish was realized indirectly. It happened that koalas were needed for a Hollywood film about Australia, and the New South Wales government agreed to lend four animals on condition that they were kept at San Diego Zoo. After the film was finished, the koalas were presented to the Zoo as a gift from the children of Australia to the children of San Diego. In time the last of these animals died, but Sir Edward Hallstrom, Director of Taronga Zoo in Sydney and Honorary Vice-President of the San Diego Zoological Society, arranged for six koalas to be sent to the USA, three to San Diego and three to San Francisco. When the trio arrived in San Diego Zoo, one of the two females was found to have a young one in her pouch. Then in 1960 the Zoo announced the arrival of the first koala to be conceived and born outside Australia. More youngsters were born in the following years and in 1976, to mark the bicentenary of the United States, the people of Australia presented six koalas to join those already at the Zoo. The thriving colony at San Diego is the only one outside Australia.

That San Diego Zoo is able to provide koalas with their natural diet of eucalyptus leaves, is due to another aspect of Dr Wegeforth's zoo plan. He realized that trees and shrubs would not only enhance the grounds but that they could also provide food for the animals. Thousands of trees were planted, many by Dr Harry himself, including exotic rarities which he collected on his travels round the world, and which thrived in their new home. As a result of this policy, many of the Zoo's animals can be fed on the same types of plants which provide their diet in the wild. Over the years the botanical collection has been enriched, and it now contains some 2,500 different species, many of them rare and threatened in their natural habitats.

Lest the opposing politics of different administrations might adversely affect the Zoo's future, Dr Wegeforth won the Society's right to run it without interference from the city. After ten years of campaigning, he also persuaded the civic authorities to provide the Zoo with an annual budget to help its upkeep. He was much aided in his negotiations by Mrs Belle Benchley, who had joined the staff in 1925 as a relief bookkeeper, a post from which she moved, by way of public relations, to become Executive Secretary. When Dr Harry died in 1941 it was she who continued his work in assuring the survival of the San Diego Zoo.

Because San Diego is a naval base, the entry of the United States into the Second World War meant a great increase in the population. As a result, zoo attendances went up. This not only helped financially but also brought the Zoo's existence to the attention of people from many parts of the country. Despite the war, the Zoo continued to rear animals, and when peace returned it was in a position to help many depleted and war-damaged European zoos to restock. It was also able to resume the guided bus tours which had originated, many years earlier, as part of its educational work.

Soon after the foundation of the Society, Dr Thompson, a member of the Board of Directors, wanted the Zoo to provide some sort of public instruction. He decided to give Saturday afternoon lectures in Park Boulevard, where the animals were then kept, and their success encouraged the Society to organize an extensive education programme. But Dr Wegeforth soon discovered that, although schoolteachers brought their classes to the Zoo, many children did not attend because they could not afford the bus fare. The answer was to introduce a free bus service, and in 1926 Miss Anna Zimmerman donated the money to buy two vehicles whose drivers soon became competent guide lecturers for the children they took on tours round the Zoo. Never one to miss an opportunity, Dr Harry realized the potential of this service, and when the buses were not needed for the schools they carried fare-paying zoo visitors. The Society now owns a large fleet of buses which every year carries thousands of people on forty-minute lecture tours of the gardens.

The Zoo's education courses are organized in cooperation with city and county schools, and summer schools are run at the zoo in the long vacation. In 1960 a summer school was started for handicapped children, for whom many special projects have since been designed. In 1966 a building, named the Elmer C. Otto Center after its donor, was opened. It houses a library, several Zoo departments, including education, and contains an auditorium which can be converted to separate lecture rooms.

In its formative days, the Zoo was greatly aided by newspaper publicity, in particular by the *San Diego Sun* which featured a weekly column entitled 'Zoonooz'. These articles, which began in 1924, contained light-hearted comments on the Zoo's doings, and two years later Dr Wegeforth started a magazine under the same name for the Society's members. Since then it has appeared regularly, and contains well-illustrated information on all aspects of the Society's work.

By the time Mrs Belle Benchley retired in 1953, the Zoo was over thirty years old. Many of its buildings were conversions of those that remained from the 1915–16 Exposition, and little building work had been possible during and after the war. Renovation was the priority.

Dr Charles Schroeder, who had worked in the Zoo's veterinary department in the 1930s, succeeded Mrs Benchley as Director and straightaway outlined a programme of redevelopment and construction. Many years earlier, Dr Wegeforth had officially dedicated the Zoo to the children of San Diego, so Dr Schroeder decided to begin the development with a Children's Zoo. It proved a great success and now, twenty-one years after its opening, it is being completely remodelled.

A Penguin Pavilion was opened in 1958, complete with cooled walking surfaces for the birds. On Horn and Hoof Mesa, moats replaced fences. Many enclosures were renovated and several flight cages were converted to walk-through aviaries. The plan was to create 'a zoo where bars and wire vanish and the world's largest wild animal collection appears in natural surroundings with a background of subtropical vegetation. . . . To provide unobstructed views of animals behind low-sculptured walls – or in areas which visitors can enter for almost direct contact with the animals.'

The Tropical Rain Forest Aviary typifies this approach. Built in 1936 to house birds of prey, it spans one of the many canyons in the Zoo. Now it is a walk-through aviary inhabited by hundreds of colourful tropical birds and containing a waterfall, streams, pools and masses of exotic plants to create the illusion of a jungle.

Construction continued throughout the sixties and many animals were provided with new quarters. The Great Ape grottoes are located on a canyon side, permitting a two-level design with an exercise and exhibition area above and with steps down to heated retiring quarters and sun rooms below. Near these grottoes are aviaries for exotic birds and the whole area is named Belle Benchley Plaza.

Another walk-through aviary houses over a hundred hummingbirds representing some twenty species. It has a brook and a waterfall where the birds can bathe and which also help to maintain the high humidity necessary for both hummingbirds and plants. Other birds which thrive in the Zoo are parrots and their relatives, and the breeding collection is one of the best in the world.

In the African Plains enclosure crowned cranes and Guinea fowl share quarters with gazelles and impalas. An Indian Plains enclosure contains Axis deer, nilgai, blackbuck and numerous Indian birds. The Dill Marine Bird exhibit was designed for oceanic birds, some of the most difficult species to keep in captivity, and a new Birds of Prey Aviary was opened in 1970. It is built around eucalyptus trees and contains artificial rocks and caves to provide suitable nesting sites.

To help pedestrians negotiate the steep canyons in the Zoo, two speed ramps, or moving sidewalks, were installed, and a 'skyfari' cable-car service was introduced to connect the two opposite sides of the grounds. All these developments have considerably changed the natural aspect of the site that Harry Wegeforth knew.

After a period of such activity, a few quiet years might have been expected. But that was not to be. Dr Schroeder began to look for a site where stock could be raised and where suitable conditions could be provided for the breeding of endangered species.

The city of San Diego owns about 11,000 acres (445 hectares) in the San Pasqual Valley, some 30 miles (48 kilometres) north of the town, and, on learning of the Society's plans, gladly cooperated by providing the land for a wildlife park to be owned by the city, but operated as an integral part of the Zoo in Balboa Park. 1,800 acres (728 hectares) of varying terrain resembling the East African high veldt were leased to the Society. Ground was broken in May 1969, and the San Diego Wild Animal Park was opened three years later. About one-third of the total acreage has so far been developed, most of which is given over to large areas where groups of mammals and birds live together as they would in the wild. Moats are used to separate incompatible species and fencing is kept to a minimum. The most extensive areas show animals from North, East and South Africa, and from Asia. Smaller displays are devoted to the Mongolian steppe, Arabian desert, and mountain habitats. A 5-mile (8-kilometre) monorail ride enables visitors to view the animals in these habitat groups.

About 20 acres (8 hectares) of the Park were set aside to display various animals in smaller enclosures. The area is known as Nairobi Village. The architecture is African in style and a 'Congo River Fishing Camp' dominates a large lagoon where numerous waterfowl live. Near the entrance to the village there is a Great Aviary for many birds native to East Africa, and at the opposite end is 'Tropical America'. This large canopied jungle habitat, which contains animals and plants native to Central and South America, is typical of a tropical rain forest, and visitors can walk among red brocket deer, toucans, and many other mammals and birds. Within the 'jungle' is a spacious temperature-controlled glass-walled herpetarium for reptiles and amphibians. The 'Bridge of Birds' circles the Sacred Ibis Aviary and leads to the lowland Gorillas' enclosure, which houses one of the few gorilla families in captivity. Among other attractions are three amphitheatres for various animal shows, and an Animal Care Center which serves as a temporary home for animals in need of special attention.

The Park contains a notable botanical collection of some 1,500 different species, many of which are endangered varieties. Areas are reserved for Californian and Australian plants, and a collection of conifers is being formed. A special garden has been created where partially-sighted and blind people can enjoy the scent and texture of selected plants. Nairobi Village is rich in South African flowers, and an extensive planting programme has provided the Park with much-needed tree cover.

Among the many species of endangered animals kept at the Park is a herd of Arabian oryx. These animals have always been hunted in their native lands, for not only is it considered a mark of manhood to course and outrun so fast and powerful a creature, but the flesh and hide are also supposed to possess magical and medicinal properties. The development of automatic weapons and the automobile meant that hunting became more efficient and, despite the legal protection afforded to the animals, populations were systematically destroyed.

As there seemed little chance of saving the oryx in the wild, it was decided to capture some in the hope that a breeding herd could be established. In 1962 the Fauna Preservation Society (based in London) and the Survival Service Commission of the International Union for the Conservation of Nature arranged an expedition to Aden. 'Operation Oryx', as it came to be known, succeeded in capturing three oryx, two males and a female. With the cooperation of the Arizona Zoological Society, the Shikar-Safari Club of Los Angeles, the World Wildlife Fund and the Zoological Society of London, the animals were quarantined, first in Kenya and then in Europe. They were then moved to the United States, to the Phoenix Zoo in Arizona, where the climate and environment were considered suitable for them.

The London Zoo sent a female oryx, Caroline, which had been in the Regent's Park collection since 1958, to join the captured animals. In the ten years she spent at Phoenix Zoo, Caroline gave birth to nine calves, and, at the end of 1977, her descendants numbered forty-five.

Another female was presented in September 1963 by HE Sheikh Jabir Abdullah Al Sabah of Kuwait, and in 1964 two pairs were donated by King Saud of Saudi Arabia. These animals formed the nucleus of the

World Herd of Arabian Oryx, and the breeding programme has been so successful that there are now about 120 individuals in captivity in the USA. At present the herd is composed of oryx kept in the Phoenix Zoo, Arizona, the Gladys Porter Zoo in Brownsville, Texas, and the San Diego Wild Animal Park. Another captive herd, not part of the World Herd, is in the Los Angeles Zoo.

The original herd of Arabian oryx in the San Diego Wild Animal Park arrived there from Phoenix Zoo in 1972. The six animals were maintained in breeding pens, and for several years they were not exhibited to the public. In 1977, some of the oryx in the Los Angeles herd were transferred to the Wild Animal Park where, in the same year, the Arabian Desert enclosure was opened to allow the public to see these rare creatures. By 1978 the number in the herd had increased to forty-two and a few of these animals have recently been sent to a desert reserve in Jordan. If the experiment is successful, more oryx will be sent to join them in the hope of re-establishing the animals in their homelands.

At present, Ministry of Agriculture regulations prohibit the importation to Britain of ungulates from North America, but some oryx from the World Herd in the USA are to be sent to four zoos in Europe, where import restrictions do not apply. Young born to these animals will then be offered to the Zoological Society of London, the donor of the female whose fertility played so large a part in the multiplication of the original group.

Apart from the Arabian oryx, other rare hoofed animals in the San Diego Wild Animal Park include breeding groups of addax, Eld's deer, Cretan goats and slender-horned gazelles. Among the carnivores in the Park are cheetahs and Bengal tigers. Two groups of cheetahs are kept and so far seven litters have been born.

The years following the opening of the Wild Animal Park were also a busy time for the Zoo. Many new collections were created or renewed, including 'Cascade Canyon', designed to display birds and semi-aquatic mammals. It stimulates a forest glen with waterfalls and brook, providing an attractive setting for waterfowl, crowned cranes and sitatunga. Another new major exhibit is the Flight Aviary built for bald eagles, which contains a stream where the birds can fish.

An important development was the establishment of a conservation programme for lemurs, rare small mammals from Madagascar. The Malagasy government allowed specimens to be collected for the Society, and in 1975 the animals were put on display in a large new enclosure in the Zoo. The breeding record has been good and it is hoped that it will eventually be possible to reintroduce some of these lemurs to national parks in Madagascar.

Other animals of particular interest in the Zoo are the okapis and bongos and a varied primate collection which includes breeding groups of pygmy chimpanzees and Douc langurs.

Zoo plans for the future include the redesigning of the Reptile and Amphibia Mesa, the construction of a Galapagos Island exhibit, and a new Bird and Primate Mesa with breeding centres for the study of species that do not reproduce well in captivity.

The Society's scientific activities began in 1927, with the establishment of the Scripps Hospital in the Zoo, and were further encouraged in 1940 when the Ellen Browning Scripps Foundation Fellowships for the study of animal health and behaviour were set up. After the war, the Foundation made a grant to the Hospital, which helped greatly in the development of research into pathology, parasitology and other aspects of animal care. A Pathology Department was established in 1964, and in 1976 an endocrine laboratory was set up at the Hospital. The following year a clinical wing, the Jennings Center for Zoological Medicine, was added to the Scripps Hospital. The new Center is excellently equipped and has made it possible to expand the programme to train veterinarians in 'zoo medicine'. In addition there is a satellite medical centre in the Wild Animal Park where, in 1972, a Wild Animal Station for Medical Care and Research was donated by Mrs Jerene Harnish, after whom it is named.

Reproductive physiology is one of the main lines of enquiry pursued in the Society's laboratories, and among various species that have been investigated is the Galapagos tortoise. These giant reptiles are native to the Galapagos Islands, off Ecuador, where their numbers have been seriously depleted as a result of the depredations of man and introduced animals. Many young have now been successfully reared at the Zoo, and the Society also contributes funds towards a conservation programme at the Darwin Research Station on the Galapagos. Other research projects are concerned with genetics, and funds are being sought to establish a virology laboratory.

In Dr Wegeforth's time, the policy of the San Diego Zoo was to display a notable group of rare animals, rather than to have a large number of any one species. Today the Society concentrates on establishing breeding groups of animals, its aim being to achieve a completely self-sustaining animal population by 1990. Between 1970 and 1977, more than 4,000 births and hatchings occurred in the Wild Animal Park alone, many among the 27 endangered species kept there.

SAN DIEGO

High standards of animal health, success in breeding, the provision of near-natural conditions for many animals and extensive research and education programmes combine to place the San Diego Zoo and the San Diego Wild Animal Park in the front ranks of the world's zoos.

PRAGUE

Zoo Praha

Colin Rawlins

Had it not been for those accidents of geography and history which made Prague the capital of Bohemia – a country which jutted as a Slav salient into the mainly Germanic heartland of Europe, and which despite an underlying leaning towards Protestantism was part of the Habsburg Holy Roman Empire – the Prague Zoo, rather than the Vienna Zoo which was founded in 1752, might well have been counted as the oldest of all existing zoos. Bohemian kings were keeping lions in Prague Castle as early as the thirteenth century. Charles IV, King of Bohemia from 1346 to 1378, not only founded the famous Charles University of Prague, but at the other end of the cultural scale established the post of *custos leonum* – keeper of the lions and other dangerous animals. By the sixteenth and early seventeenth centuries the Prague collection of wild animals was probably unequalled in Europe.

In 1534 King Ferdinand I, who had been elected to the Bohemian throne in 1526 and who later also became Holy Roman Emperor, took over some vineyards as land for a zoological garden. Maximilian II and Rudolph II, his successors on both thrones, subsequently built on this land a menagerie which became the envy of the royal houses of Europe. Maximilian later stocked a park near his Prague castle with deer, bison and other hoofed animals. On his journey from Spain to Vienna in 1552 he had taken with him the first elephant ever to be seen in Austria, and he had also built a Lion Court in which lions and other beasts of prey were kept. But it was in the time of Rudolph II, who reigned from 1577 to 1602 and who is said to have been indifferent to religious controversy and matters of state, and yet a great patron of the arts, that the Prague Menagerie, which was already developing along the lines of a modern zoo, reached the zenith of its fame. Apart from lions and tigers, the collection is known to have included cheetahs and an orang-utan.

The bones of a dodo (*Raphus cucullatus*), which were found in Prague towards the end of the nineteenth century, may well have been those of a live bird, then as much a rarity as giant pandas are today (but now extinct), that had been brought to Prague from Mauritius. In about 1600 the painter Hoefnagel of Prague made an oil painting of a young male dodo which was also probably part of the royal collection. The renowned English traveller Fynes Moryson, who visited Prague in 1592, reported with admiration and wonder at seeing twelve lions. Emperor Rudolph took a personal interest in his animals; it is said that his favourite, a big lion, died on the same day in 1602 that he did.

But with his death, Prague began to cede its political position as the centre of the Holy Roman Empire. In May 1618 the terrible Thirty Years War opened with the famous 'Defenestration of Prague', when three imperial officials were thrown from a window of that same castle near which the royal menagerie was kept. The accession as Emperor of the Jesuit Ferdinand II, who had pledged himself to crush Bohemian Protestantism, hastened the decline, and the centre of the Empire moved inexorably to Vienna. So too did the menagerie, and little more than a century later, with most of the animals from Prague now housed there, the Vienna Zoo was opened to the public in the garden of the royal palace of Schönbrunn.

Prague was no more than one of the provincial capitals of the Austro-Hungarian Empire when new zoological gardens were being opened during the nineteenth century in many of the big cities of Europe and America. Those Prague citizens who sought to start a zoo ran into many difficulties, and it was not until Czechoslovakia became independent at the end of the First World War that a zoo was established, mainly as a result of the determination of one man, Dr Jiri Janda. In 1924 he succeeded in forming a Prague Zoological Garden Society. Two years later a formal administrative body in the shape of a 'cooperative' in which both the State and city governments played a part, was set up to manage the new Zoo under his direction. His ambition was finally realized when the Zoo, the Zoologicka Zahrada Praha – the Zoological Garden of Prague – was opened to the public on 28 September 1931.

Dr Janda wanted to lay out his zoo as a wild park, and an appropriate site was found on the north of the Vltava River, just 2½ miles (4 kilometres) from the heart of the city. Only 20 acres (8 hectares) when it opened, today it covers 150 acres (61 hectares) of a rocky wooded hillside overlooking the wide river. It is an almost perfect site for a national zoo which has education and recreation as its purpose.

Very few animals were available when the new zoo opened in 1931, but their numbers slowly increased with gifts and with purchases from circuses, which at that time were being forced to reduce their stocks because of the prevailing economic depression. One important gift was a pair of Przewalski's horses from the Czech government. Although at that time the government did very little else to support the new zoo, this gift of a breeding pair of a very rare species was later to prove to be the Zoo's most important animal acquisition.

Dr Janda was over sixty when the Zoo opened in 1931, and he died in 1939. During and immediately after the war his two successors, Dr Stanek and Dr Vlasak, kept the Zoo going in spite of great difficulties, with shortages of money and food, and with no prospect of replacing or adding to the animal stocks. Dr Vlasak even managed in 1942 to rear a polar bear cub by using the milk of a cow which had to be kept secretly in the Zoo.

Recovery was rapid after the war, and the Zoo's expansion, for which Dr Janda had planned and worked so hard, was soon under way. More land was acquired, much of it in a remarkably natural state for an area so close to a great city. The financial constraints of the early years disappeared when what had started as a private institution became incorporated as a cultural service of the city of Prague. Dr Purkyne was appointed Director, and by its twenty-fifth anniversary in 1956, Prague Zoo had achieved international standing as a zoo of quality and scientific importance.

Dr Purkyne was succeeded in 1959 by Zdenek Veselovsky, a young naturalist of rising reputation, and in the same year he organized a symposium on the conservation of Przewalski's horse. Prague Zoo then became responsible for keeping the international studbook of this species, which now breeds freely in several zoos.

Equus przewalski, sometimes also known as the Asiatic or Mongolian wild horse, was named after the nineteenth-century Russian naturalist and explorer Nikolai Przewalski. Originally a native of Siberia and Mongolia, the animal is either extinct in the wild or exists only in very small numbers. Its ultimate preservation, therefore, is dependent upon the care and increase of the stocks kept in zoos. The studbook, analagous to records kept by domestic animal breeders, contains the details of the pedigree of all Przewalski's horses in captivity. As the original breeding stock was small, and as some crosses with domestic horses were thought to have occurred, it became important that all available data should be posted in one place. The work of the Prague studbook keeper, complemented by the knowledge disseminated through three international symposia, forms the basis of the scientific management of the species throughout the world.

The first Przewalski's horses were introduced from the wild in the late nineteenth century, first to the famous Askanya Nova estate in southern Russia, and then to Halle in Germany and to Woburn Park, the seat of the Duke of Bedford, in Britain. Until the 1950s the number of animals in these collections and in other zoos rarely exceeded 40; by 1958 the figure had risen to 50; by 1964 to 100; and by 1975 to 250.

The pair of Przewalski's horses which Prague Zoo received in 1932 had belonged to the Zootechnical Institute of the Prague Agricultural High School. The stallion died soon afterwards, but another male was brought in from the Washington Zoo in the USA and it is his progeny which formed the basis of the present-day Prague stud. By the end of 1977, 135 Przewalski's horses had been born in the Prague Zoo, and of these more than 50 were sent to other zoos to encourage further breeding and to improve the stock by crossings of known parentage.

Prague Zoo today houses some 2,500 animals, including about 400 fish. There are about 700 mammals belonging to 150 species, over 1,100 birds of 250 species, and 120 reptiles of 50 species. There is a preponderance of Asiatic animals, the result of the strong new contacts built up with the East in the post-war years. Australian and South American species are also well represented, the former reflecting a special interest of the present Director, whose zoological expeditions to Australia have been recorded in a number of books and articles.

Among the Zoo's notable breeding successes are two South American animals. The first is the maned wolf, of which thirty-eight had been bred in Prague by the end of 1977. The maned wolf is a beautiful long-limbed creature of the Pampas, which is not often bred elsewhere in captivity. The second is the two-toed sloth, which is as slow as the maned wolf is fast. Prague has successfully reared several.

Tigers have always been a feature of the Prague Zoo. The rare Siberian sub-species has done particularly well, and thirty-five cubs have been reared. A group of the even rarer Sumatran race of tiger is also breeding. Several other members of the cat family, most of which have bred, can be seen in the Zoo. Notable among these are the small ocelot from South America, the snow leopard from the mountains of Central Asia, and the clouded leopard from south-east Asia. The cheetah, a species which did not breed in captivity until a little more than a decade ago, has been Prague's most recent success, with twenty-seven born since a first litter in 1972. This achievement has been exceeded by only a few other zoos – among them the Zoological Society of London's Whipsnade Park and the cheetah breeding establishments of the San Diego and Pretoria Zoos. The success Prague has had with cheetahs may well be partly due to the relatively open and spacious accommodation which the Zoo provides for its animals. Successful cheetah breeding in captivity needs paddocks or enclosures in which the animals can run around and, as has been suggested, from which they can have an unrestricted view.

The stars of the Prague collections, the Przewalski's horses, share with other herds of hoofed animals the lush enclosures on the hillside and hilltop. Here can be seen African antelope, such as the nyala from southern Africa and its close relative the sitatunga, inhabitant of the African swamps and marshes, as well as the very different scimitar-horned oryx, a desert animal from the southern fringe of the Sahara. The steep slopes are the right background for mountain goats and

Prague keeps and breeds the rare markhor and goral or 'Himalayan chamois', both from the mountains of Asia.

Asiatic wild asses or onagers are animals whose natural habitats have been seriously reduced and affected by human encroachment. Breeding them in captivity (there are about four hundred in zoos) is one way of ensuring that they do not die out altogether. Prague has a good collection of the Turkmenian sub-species, generally called the kulan, and bred twenty-three up to the end of 1977.

As is well known, the European bison, or wisent, a native of the forests of Eastern Europe, would already have become extinct if it had not been preserved in zoos. It was all but exterminated in the wild during the Second World War, but fortunately a few members of the species still survived under captive conditions, particularly in central and eastern Europe, to form the nucleus of a breeding stock. Some were then returned to the wild in Poland and Russia, where the species is now securely re-established. Prague Zoo has helped in this classic piece of wildlife conservation; up to the end of 1977 it had bred and reared twenty-seven wisent.

Dr Janda wanted the Prague Zoo to be first and foremost a park and consequently there are few major animal buildings or obtrusive installations. There are pavilions for apes and monkeys and for other small mammals, and a large house for the bigger carnivores such as lions and tigers (this also houses some reptiles, including crocodiles). The most imposing building of all was opened in 1970 to house elephants, rhinoceroses, hippopotamuses and tapirs. Nearly all other animal houses have been simply constructed to allow good management without detracting from the natural aspect of their assorted outside enclosures.

The topographic and scenic advantages of the steep slopes of the Zoo have been exploited not only in the maintenance of various mammals which can tolerate the sometimes harsh climate of central Europe, but also in the siting of a spacious and lofty aviary for birds of prey against an almost vertical cliff-face. In 1937 Andean condors, a species which rarely breeds in captivity, were reared in this aviary. In 1969 the structure was damaged in an extremely heavy snowfall. It was then replaced by another, which is every bit as fine.

Offices and service buildings are confined to the eastern end of the Zoo where the public entrance is situated. A recent development has been a chairlift to take visitors from the more built-up lower level of the Zoo to the open area of herd-animal paddocks on the upper slopes of the hill.

Public education and zoological research have always been part of the work of the Prague Zoo, and Dr Veselovsky is an Associate Professor of the Charles University. There is a comprehensive and excellent system of information notices and tape recordings at various points in the Zoo through which visitors are informed about zoological matters. One interesting educational innovation is what are known as the Prague Zoo's 'days of questions and answers', when members of the public are invited to watch films and then discuss the subject matter. During the summer, students from the Natural Sciences Faculty of the University are employed to act as free guides for organized groups visiting the Zoo.

There is also a programme designed to help the Prague schools in their teaching of biology. Primary-school children are given talks on animals at the Zoo. Secondary school pupils are instructed in relation to their biology syllabus, while advanced pupils preparing for university entrance in the Faculty of Natural Sciences are given the opportunity to work on simple scientific projects. Seminars on the use of the Zoo by biology teachers, and on animal husbandry and the biology of wild animals, are also regularly organized.

Several natural history and zoology clubs, the most important of which is the Club of Zoo Friends, make use of the Zoo's educational facilities. Competitions connected with animal subjects are arranged for members of this and other clubs. The Zoo publishes a combined magazine and annual report, *Gazella*, in which appears the results of studies encouraged by Dr Veselovsky, who is particularly interested in the study of animal behaviour.

About 700,000 visitors come to the Zoo each year; as Prague's population is little more than a million, this is a remarkable figure. Communications in the city itself are continually being improved and this is expected to bring in even more visitors.

Czechoslovakia has a relatively large number of zoos, including some such as the Herd Animal Centre at Dvur Kralove in the north. But as the national zoo it is Prague's responsibility to advise the government on zoo matters and to coordinate statistical and other information from all Czech zoos for the government's use. Among the Slav nations, the Prague Zoo plays a leading role as a fine example of the good modern zoo, well managed and with a clear scientific purpose.

EAST BERLIN

Tierpark Berlin

Colin Rawlins

The division of Germany at the end of the Second World War left three zoos in the eastern zone: those of Leipzig, Dresden and Halle. The little that remained of the Berlin Zoo was in the western sector of the city, and the people in the eastern part, more numerous than the population of any other East German city, were soon canvassing the idea of having their own zoological garden.

They did not have long to wait. The German Democratic Republic received its first formal constitution in 1949. Only five years later the government decreed that its capital city, East Berlin, should also have a zoo. The site chosen was in a large park called Friedrichsfelde towards the east of the city, and Dr Heinrich Dathe, then Assistant Director of the Leipzig Zoo but with experience in zoo management going back to pre-war days, was put in charge.

With the growth of the city over the past thirty years, the Friedrichsfelde is no longer on the fringe but at the centre of East Berlin. It covers nearly 400 acres (162 hectares) and comprises heavily wooded parkland on relatively flat and easily managed ground which is dominated by a stately house with formal gardens. The property has a long history. Soon after the first recorded mention of the town of Berlin, there is a reference to a homestead called 'Rosenfelde', which for several hundred years was owned by the various families which in succession became the Electors of Brandenburg and finally, through the Hohenzollerns, Kings of Prussia. In the seventeenth century Rosenfelde came into the possession of a Dutch merchant, Benjamin Raule, who had come to Berlin to help the Great Elector, Friedrich Wilhelm, create a navy and develop overseas trade. Raule built a summer pavilion in the Dutch style and laid out gardens in the contemporary formal baroque manner. Some 150 years later, in the reign of the Elector Friedrich III, the estate reverted to royal ownership and was renamed Friedrichsfelde. The undeveloped land leading from the residence and its formal gardens was then laid out as a park by P. J. Lenné, a renowned landscape architect, who was responsible for the landscaping of the Prussian royal residence of Potsdam as well as for the new zoo in the Tiergarten to the west of Berlin. In 1861 the whole estate was bought by a recently ennobled family, the Treskows, who kept it until 1945, when it became the property of the new State.

Dr Dathe realized that his appointment as the first director of a new national zoo gave him an unrivalled opportunity to design an entirely modern zoo of major proportions. Moreover, the site did not need traditional barriers to restrain animals. The Schloss and its formal gardens could be preserved as the base from which the visitor would start his tour of the Zoo. The main public entrance could be in this area, and the exhibits and roads could be laid out in such a way as to take visitors on a circular tour through open displays including, at the south-eastern end, an exceptionally large multi-purpose building for animals needing protection from the weather.

The speed with which decisions about Tierpark Berlin, as the new zoo was to become known, were taken and then implemented must have been envied by other zoo directors. By October 1954 a layout plan was ready and a month later a foundation stone was laid. Dr Dathe immediately took up residence in temporary offices in the Schloss, and in no time construction work was started. A year after the authorities had decided to have a zoo, part of it was opened to the public.

Schloss Friedrichsfelde is now a 'protected' building, and is being restored to its original design. New water-courses enhance the elegance of the replanted formal gardens, and the ornamental ponds have been stocked with colonies of pelicans. From the Schloss there is a clear view to the south-east for more than 1,000 yards (900 metres) along an avenue flanked with oaks and lined with rhododendrons and azaleas. In 1973 a second public entrance was built, leading on to the southern end of the avenue.

One striking thing about the Tierpark Berlin is its vast size in the centre of a major city. Moats, which look like forest pools, keep the public out of open glades that have been transformed into grazing grounds for hoofed animals. Most of the barriers, on the service side of the enclosures, are well hidden by abundant tree cover. The whole result is a magnificent series of exhibits of antelope, deer, bison and other hoofed animals.

There are attractive enclosures for other animals that can be kept in the open. The polar bear enclosure must be one of the best of its kind. It has a naturalistic rock background, made of blocks of granite from the ruins of the pre-war Reichsbank, and its pool covers half an acre (fifth hectare), being 100 yards (90 metres) long at one point. The enclosures for the other bears are equally attractive. Another fine display is the island for gibbons, which is provided with an appropriate South-east Asian look by a climbing-frame of bamboos set against a planted background.

Possibly the largest building in the world for zoo animals is the Alfred Brehm House, which took six-and-a-half years to build and which was opened in June 1963. Named after the nineteenth-century German naturalist, it is almost a zoo in itself. The house covers 1½ acres (5,300 square metres) and is over 100 yards (90 metres) long, with a maximum height of 50 feet (15 metres). Its imposing entrance is flanked by two giant bronze lions, once the sentinels of the monument at the front of the former German Imperial Palace in Berlin.

The central part of the Alfred Brehm House is the Tropical Hall, planted with jungle-type vegetation, and containing free-flying birds and fruit bats. Temperature and humidity are controlled throughout the year. At one end of the building lions and tigers are confined in four large dens backed by natural rock and separated from the public by wide moats. Two wings of the house comprise cage units, each of which consists of an outside enclosure, an isolation den and an inside den. The outer enclosures are separated from each other by shrubs, and the inside dens by sheets of glass. Small mammals, including small cats and marmosets, birds, reptiles, fish and insects are exhibited in the different units.

In 1965 the Alfred Brehm House was extended to the north-east by the addition of a flight aviary for birds of prey. The aviary, which again is one of the largest exhibits of its kind in the world, being about 70 yards (64 metres) long and 30 feet (9 metres) high, provides enough room for even large birds of prey to display their soaring abilities.

The efficiency which marked the building of the Zoo has also characterized the acquisition of animals. The Zoo's animal collections are truly spectacular, both in number and variety. At the opening in 1955 there were 2,000 animals of 300 species, mainly birds and mammals. By the end of the 1960s, the figures had grown to 5,000 animals of more than 850 species. With the amount of space at its disposal, it is not surprising that hoofed animals have been made a feature. There are some fifteen species and sub-species of deer on exhibition, most of them breeding regularly, including such rare ones as musk deer, barasingha or Indian swamp deer, brow-antlered deer and Indo-China Sika deer – all natives of South-east Asia. Rare sub-species of red deer, muntjac or barking deer, hog deer, Père David's deer and roe deer are also kept.

Antelopes are also well represented, with breeding groups of scimitar-horned oryx and addax from the deserts of Africa and Arabia, bontebok and nyala from southern Africa, blackbuck and nilgai, sometimes called the blue antelope, from India, and a large herd of the timid saiga from the Asian steppes. The East Berlin Zoo, like other zoos in Germany and Eastern Europe, breeds the once nearly extinct wisent, or European bison. Over fifty individuals of this species have been bred in the past twenty years, and the Zoo now has one of the biggest captive herds in the world.

It also keeps a large group of the Asiatic wild ass, including the largest captive colony of the kulan sub-species from the Turkmenian region of central Asia. In all, there are now about 350 individuals of this endangered species in zoos, and the steady increase in their numbers over recent years is a remarkable illustration of what zoos can do to conserve disappearing animals. The Zoo maintains the international studbook, not only for Asiatic wild asses but also for the two species of African wild ass.

East Berlin is also notable for its large and small cats in the Alfred Brehm House. A fine breeding group of the rare Indian lion, three subspecies of tiger and five different races of leopards are the main representatives of the big cats. There are also snow leopards, clouded leopards, pumas and jaguars. European wildcats, the leopard cat from the Indian jungle and Geoffroy's cat from the Brazilian forests are some of the smaller cats that are exhibited.

All seven species of bear known today are to be seen in their spectacular granite-backed enclosures. The large Kamchatka form of brown bear, polar bears, the small sun bears from Malaysia, Asiatic black bears and the rare spectacled bears from South America – the only species to be found in the southern hemisphere – have all bred in the East Berlin Zoo. The first sun bear birth was in 1961, and there have been births regularly since then.

The bird collection in East Berlin, numbering 2,000, is as varied as are the Zoo's mammals. Waterfowl are to be seen in many watercourses throughout the Park; tropical birds in the Alfred Brehm House; and birds of prey in the Great Aviary. The tawny eagle, black vulture, Himalayan griffon, long-legged buzzard, and the bald eagle have bred in the large aviary. The breeding of the bald eagle is notable. The only other captive breeding of this eagle, the national emblem of the United States, has occurred in its national home. Six different species of pelican live on the ornamental pond in the formal gardens of Schloss Friedrichsfelde. No other zoo in the world has such a remarkable show of these fascinating creatures, and the Zoo's breeding record is unique. All species – the white, Dalmatian, pink-backed, brown, Philippine and American white – have bred. The hatching of brown pelicans in 1965 was the first recorded captive breeding of this species in Europe. The white pelican has been breeding regularly since 1961.

The East Berlin Zoo was founded to provide not only an educational and cultural service for the people of the German Democratic Republic, but also a centre of scientific research. As early as 1958 a unit for vertebrate animal research was established in the Zoo, under the auspices of the Science Academy (Akademie der Wissenschaften) of the German Democratic Republic and with close ties with the veterinary faculty of the Humboldt University in Berlin. The twenty-five professionals who make up the staff of the unit are exclusively occupied with zoological and pathological research, and there are excellent modern laboratories and offices to support this work. The Zoo's staff have been and still are responsible for the organization of the annual meetings held in Europe to discuss the diseases of zoo animals.

The East Berlin Zoo publishes the international journal, *Der Zoologische Garten*, started more than a hundred years ago by Dr Max Schmidt, the famous Director of the Frankfurt Zoo. It is the most important professional journal of its kind in the German language, covering a wide range of topics, including those related to the management of animals in captivity. Another international journal which the Zoo publishes is the *Beiträge zur Vogelkunde* (*Contributions to the Science of Birds*). A more popular magazine called *Milu*, the Chinese name for the Père David's deer, is issued once a year. It consists of items about the Zoo, its staff and its animals, as well as more professional articles.

Education is also an important part of the Zoo's activities, with a curator and four qualified teachers in an Educational Department whose purpose is to give guidance to visiting teachers from Berlin schools on the best way of using the Zoo for the teaching of biology. In 1963 the Zoo founded a Youth Club, with a present membership of about 450 young people.

Tierpark Berlin has provided the people of Berlin with a place of rest, enjoyment and education where, in a beautiful and ample setting, the cares of city life can be forgotten and the mind stimulated and broadened by one of the greatest collections of wild animals anywhere. It has taken only twenty-five years to develop the largest urban zoo in the world, with one of the most comprehensive collections, and with a powerful emphasis on scientific and educational work.

TORONTO

Metro Toronto Zoo

John Toovey

In 1974 the city of Toronto opened a zoo that covers more than a square mile (287 hectares), and which in its first year had a million visitors. It is by far the largest civic zoo to be built in recent times, and is laid out so that visitors can see animals in the zoogeographical (as opposed to the continental) zones in which they are found in nature, i.e. the Eurasian, North American, South American, African, Indo-Malayan, and Australian regions. In each region, the visitor first passes through a pavilion where information about the animals and plants of the region concerned is displayed. These pavilions, of which four have already been built, house the more delicate animals, and open on to outdoor paddocks and pools where the hardier animals live.

The simplest way to give an impression of the whole Zoo is to describe the 'Round the World Tour', one of the several routes which have been planned for visitors. The most comprehensive, it takes at least three hours, whilst even the shortest recommended walk takes over two.

The visitor enters the Zoo on the eastern side, and moves along a path which encircles 80 acres (32 hectares) of mature deciduous woodland. Around this woodland are areas representing the six different zoogeographical regions. Walking north along the path, the visitor comes first to the Eurasian area. The pavilion here houses only small mammals, birds and reptiles. Much of the building is designed as a walk-through aviary without barriers, the birds being contained by bamboo curtains. In warm weather, a glass partition can be opened to give access to an outdoor extension of the aviary. With a mainly Mediterranean setting, the plants used vary from desert to sub-tropical species, including papyrus, lime, fig, banana, jasmine and palms from North Africa.

On leaving the pavilion, the visitor is directed along the 'Tiger Trail' which leads into the 50 acres (20 hectares) set aside for Eurasian animals. This area contains a wide selection of animals from Europe, Russia, Africa north of the Sahara, the Middle East, China and Japan – the Old World land mass. With a range of species that extends from tigers, adapted to very cold environments, to oryx from hot, arid deserts, there is ample opportunity to illustrate many different habitats and ways of life. Originally flat, the site has been landscaped to simulate Eurasian topography, with lakes, steppes and rocky deserts surrounded by hills which represent mountains. Animals exhibited here include Siberian tiger, Chinese leopard, Caucasian tur, Père David's deer, scimitar-horned oryx, reindeer, yak, European bison, Arabian camel, Barbary sheep, Japanese macaque, cranes and storks.

Turning west, the visitor leaves the Old World for the New, beginning with the North American area, where he first passes an open-air polar bear enclosure with underwater facilities which allow him to see the animals' graceful movements when swimming and diving.

The North American Pavilion, like its Eurasian counterpart, houses only the smaller animals, the larger species being kept outside. The building is set into the ground, its roof thus being used both as a continuation of the main path and as a platform for viewing otters, beavers and muskrats in their outdoor ponds. These animals can be seen underwater from inside the building, while others, including twilight and nocturnal animals, are provided with swamp, desert or prairie settings. Among those to be seen are prairie dogs, nine-banded armadillos, skunks, turkey vultures, flying squirrels, rattlesnakes and cacomistles, as well as alligators and American crocodiles. The planting represents two main areas: the Florida Everglades, with swamp grass, cabbage palms, gumbo limbo trees, mahogany and cannas; and the South-western Desert, with cacti, agaves and yucca plants.

The World Tour now continues along the west side of the woodland to reach first the African, and then the Indo-Malayan Pavilions. These two buildings are similar in concept and structure, housing mainly species from tropical jungles or rain forest. The artificial environment of the buildings can therefore be consistent throughout, while each is landscaped as an entity. Barriers are unobtrusive, and the skilful handling of space and materials goes a long way to create the illusion of jungle. The hyperbolic paraboloid roofs, made of large areas of glass alternated with solid wood, are high enough to enable the growth of trees which create patches of light and shade similar to the contrasts found in nature. With an irregular floor surface and public walk-ways at varying levels, the animals are seen from many different viewpoints.

The African Pavilion covers an area of over 1 acre (0.5 hectare) and the roof rises to over 40 feet (12 metres). The interior is divided into two – the southern half representing wooded swamp, and the northern half dense African jungle. The height of the building makes it possible to grow many trees and vines, such as traveller's and triangular palm, screw pine, tamarind and kapok trees, fig, banana and clerodendron. Among the 'jungle' animals are gorillas and mandrills, as well as bongo (one of the larger antelopes), crowned cranes and the African dwarf crocodile. The 'swamp' area contains pygmy hippos, sitatunga, De Brazza's monkeys, Congo buffalos, black-footed penguins and hornbills. Beneath the elevated walkways are tanks containing reptiles and fish.

From the African Pavilion the 'Lion Trail' leads to the 50-acre (20-hectare) Savannah area where a mixed-species collection includes white rhinos, Grant's zebras, greater kudus and several varieties of ground-living birds. Separate enclosures are provided for typical African species such as lions, springboks, addax, cheetahs, gnus, giraffes, blesboks, baboons, lechwes, hippos and elephants. As most of these animals cannot adapt to the intense cold of a Canadian winter, they are also provided with indoor accommodation.

Since few native African plants can tolerate the Canadian climate, an African landscape had to be simulated. What is now the Savannah area was once an apple orchard, and many of its trees have been preserved and pruned to resemble African acacias. New trees suited to Canadian conditions, among them pseudacacia, hawthorn, ash, cotoneaster and ailanthus, were selected for their African characteristics. The visitor has the impression of being in an African landscape as he wanders past these paddocks towards the woods which act as a screen to the next zoogeographical area.

In the Indo-Malayan Pavilion – just under 1 acre (0.35 hectare) in extent and with its roof reaching to a height of 60 feet (18 metres) – the visitor is suddenly transported into a Sumatran rain forest where the lush vegetation includes palm and mango, rubber and umbrella trees, as well as crotons, hibiscus, ginger and epiphytic orchids and, as a dramatic centrepiece, a magnificent banyan tree. There are a number of free-flying birds, including a pair of southern pied hornbills; aquaria containing fish typical of the region, and reptiles, including gharials, mangrove snakes and Tokay geckos. The main attractions are the orang-utans and the gibbons, the latter sharing their enclosure with Sarus cranes and Malayan tapirs.

Leaving the Indo-Malayan Pavilion and turning east, the path leads to a smaller but fascinating exhibit. the 'Village Edge Pavilion', housing plants and animals that co-exist with man on the fringes of Oriental villages. Turning north, the visitor then arrives back at the Zoo's entrance.

The Australian and South American regions, in the east and north-west of the Zoo, are still relatively undeveloped and the appropriate pavilions have yet to be built.

As well as exotic animals and plants, the Zoo is justly proud of its 'Canadian Animal Domain', where Canadians may see, most of them for the first time, their native fauna. Lying to the south-west of the plateau on which the main buildings and entrance are sited, and set against the spectacular 410-acre (166-hectare) Rouge River valley, this entirely outdoor area is an important part of the Zoo's conservation effort. The Domain is open all the year round so that visitors can enjoy the changing colours of the splendid woodland, in which the original pine, ash, cedar, birch, poplar and sycamore have been preserved. New planting includes cranberry, willow, sumach and alder. Each of the paddocks contains an area of high ground to which the animals can retreat in case of flooding. Apart from the unobtrusive mesh fences, necessary to keep incompatible species apart, a small electrically powered rubber-tyred train is the only man-made intrusion in the valley.

For those who do not wish to walk, the free twenty-five minute train ride to the Domain is an excellent alternative to the tours, crossing as it does a large part of the Zoo. It can be enjoyed in heated or air-conditioned comfort, according to the season, and the train, which can carry about 150 passengers, moves slowly enough to allow plenty of time to see and photograph the animals which have been introduced into the Canadian Animal Domain. These include white-tailed deer, wapiti, timber and Arctic wolves, wood bison, Dall's sheep and many waterfowl. Indigenous birds and animals – racoons, muskrats, skunks, otters and minks – can also occasionally be seen. A break in the journey allows the visitor to see the grizzly bear and coyote enclosures, and to these will be added an enclosure for Canadian cats.

This is the Metro Toronto Zoo as seen today. But at the beginning of the 1970s, none of this complex organization existed. How does a new zoo grow from nothing? And what are the problems?

Toronto already had a zoo, built in 1887 at Riverdale, but by the late 1940s it had become hopelessly overcrowded and out of date. A group of local citizens then put forward a scheme for its improvement, and in 1966 Hugh Crothers, a local man, encouraged by the Metropolitan Council, formed a Zoological Society. A beautiful and varied site to the east of the city was selected, a feasibility study prepared, and in 1969 a master plan for a new zoo was drawn up. The site is crossed by the two dramatic Rouge River valleys, one of which, even at this early stage, was earmarked for free-roaming Canadian animals. In 1969 the Society was incorporated as the Metropolitan Toronto Zoological Society and in the following year the Metro Toronto Council undertook to build the Zoo. The work was supervised by a team of consultants, working with the city council and the Zoological Society. The size of the project was so great that three firms of architects were jointly employed, and this resulted in a high standard of planning and architecture throughout the Zoo, with the buildings being well designed and carefully blended into the landscape.

Work began in the autumn of 1970, and three and a half hectic years followed before the Zoo was opened. The ground had to be torn up to lay pipes and cables; complex buildings had to be erected; and the foundations of the animal and plant collections laid. One of the main priorities of the Society was the development of its membership and the provision of financial support. A membership drive, begun in 1971, was helped by visiting celebrities and by projects involving local schoolchildren. By 1975 these activities, plus various grants and donations, had raised $6 million.

Vital to the Zoo's success, of course, was the acquisition of its future inhabitants. The choice of animals was carefully planned, and then came the problem of obtaining them. A few early zoos had mounted their own expeditions to bring back ark-loads of exotic 'wild beasts'. But in these days of prohibitive costs, and with the list of endangered and rare species growing at an alarming rate, this was no longer possible. Some of the Metro Toronto Zoo's animals were obtained from the wild under licence, but most of them came from other zoos where they had been bred.

By late 1973 some 700 species were in temporary accommodation on the Zoo site, and in May 1974 the first animal, an African antelope, the bongo, was moved into its permanent new home. In 1974, the first year of operation, there were 360 births, a number of them to members of rare or endangered species.

Parallel with the acquisition of stock, plants were obtained to recreate as closely as possible the animal's natural environments, ranging from equatorial rain forest through woodland to swamp and desert. Most of the plants came from nurseries in Florida, while others were grown from seed obtained through the International Seed Exchange Programme. The collection was started early in 1973, but the final planting had to wait until the builders moved out in January 1974. Maintaining healthy growth in the tropical pavilions is dependent upon proper climatic control, with appropriate humidity, and with day and night temperatures kept within a range of 16–21°C. As with many indoor plants, the risk of disease and parasites is high, but the zoo gardener has an additional problem since the use of insecticides in or near animals might injure them. To avoid this danger, Toronto has experimented with biological control, one of the more successful ventures being the introduction of wasps to control that persistent pest, the white fly.

The outdoor landscaping was on a massive scale. 20,000 trees and shrubs of more than 75 species were planted, 200 acres (82 hectares) of grass were sown, ponds, lakes and water systems were formed, and some 2,500 tons (2,545 tonnes) of rock were used.

As in all large zoos, the public exhibits are only the tip of the iceberg, and Metro Toronto has a considerable back-up service and adminstrative organization, the latter being shared between the municipality, who own the Zoo, and the Board of Management, who operate it.

The main service area, which includes a hospital and laboratory, is sited well away to the north edge of the Zoo, but it is connected to all buildings and animal areas by a system of service routes which are separated from the public paths. Fifty maintenance men look after the vast quantity of mechanical equipment needed to produce 'natural' environments; they repair and maintain the buildings and keep everything clean. Here also are the stores and kitchens where seven or eight staff prepare food for the 3,500 or more animals in the Zoo. Over twenty full-time gardeners maintain all indoor and outdoor plants, operating from the service area where they have a large greenhouse for the propagation and keeping of stock.

With so many new animals having been introduced in the short time since 1973, the animal hospital and laboratory staff have been engaged in a heavy programme of work. Most veterinary problems can be dealt with in the Zoo, but occasionally animals are sent to Ontario Veterinary College for more complicated surgery. As well as day-to-day medical care, the staff of the hospital are involved in preventive medicine, in testing and treatment, in the establishment of well-balanced and complete diets, and in post mortem examinations. They also help with the training of keepers, and hold lectures for students and groups of visitors.

The Zoo has already made good progress in informing and educating visitors through well-considered and exceptionally well-designed graphics and literature. Volunteer guides take school parties on tours, but fulfilling the Zoo's intended educational role awaits the construction of an education centre and auditorium.

The Toronto Zoo illustrates the modern approach to zoo design and animal management, in which the trend is away from the exhibition of animals in caged groups, and towards the concept of an open 'biological park'. Through the skill and determination of many people, a complete and forward-looking professional zoo has been created in six years. The visitor benefits from an entertaining and unusual view of the many parts of our natural world, and is helped to understand the interdependent roles played by climate, geography, plants and animals. In turn, the plants and animals may benefit from man's better understanding of the consequences of pressure on the environment.

SPECIALIST COLLECTIONS

Jersey Zoo

Michael Brambell

Jurong Bird Park

Peter Olney

Clères Bird Collection

Peter Olney

Slimbridge Wildfowl Trust

Peter Olney

Arizona-Sonora Desert Museum

Peter Olney

Many small zoos, of which a few arbitrarily selected representatives are described in this chapter, have confined their attentions either to particular groups of animals, or to a particular zoogeographic region, or to habitat types. Aquaria and reptile collections have been omitted, since they enjoy a considerable literature of their own. Those who are interested will find in the *International Zoo Yearbook* a comprehensive list of all specialist collections.

The zoos which have been selected for mention in this chapter are the Jersey Zoo, owned and managed by the Jersey Wildlife Preservation Trust, and specializing in the breeding and studying of rare and endangered species; Jurong Bird Park in Singapore, a government-owned collection consisting mainly of South-east Asian birds; Parc Zoologique de Clères in France, a privately owned bird collection; the Wildfowl Trust in the United Kingdom, devoted to one family of birds; and Arizona-Sonora Desert Museum in the USA, which displays many different animal habitats found in one geographical area.

Specialist collections can provide biological information which it would be difficult or impossible to obtain elsewhere; they can help in conserving endangered species; and they can help inform the public about the diminishing fauna and flora of the world. Most specialist collections, whatever the reasons for their formation and survival, fulfil a vital function in the zoo world.

Jersey Zoological Park

Michael Brambell

The Jersey Zoo was started in 1959 by Gerald Durrell, the popular author, whose first books, *The Overloaded Ark*, *The Bafut Beagles* and *My Family and Other Animals*, revealed him as not only a lover of animals but also a writer of talent in a family whose literary prowess had already been established by Lawrence, his elder brother. In *Who's Who* Gerald states that his only professional training was that of a student keeper at Whipsnade. He spent ten years mounting and leading expeditions to collect animals for zoological gardens, and it was during this period that he first started writing the books that were to make his work and his name so well known. But with the handsome royalties that he earned from his books, and conscious of the increasing recognition of the desirability of conserving wildlife, he decided to make a home for his animals in Jersey. He started the Zoo as a private venture, but four years later transferred the management of it to a newly formed Jersey Wildlife Preservation Trust, several of whose members were professional men who had made the island their retirement home. Today the Trust, which was set up as a charity, is well established, with some six thousand members in thirty-eight countries contributing funds. The success of the Jersey Zoo derives both from its founder's genius for wise publicity, and from the dedication and skill of his staff.

The Zoo is in many ways unique. No other zoo which started as a diverse collection of animals has in so short a time and in so deliberate a way turned to keeping only endangered species. Conservation and education have for decades been the concern of all serious zoos. But by focusing on the theme of conservation, Jersey has managed to turn the limelight on to the urgent need to protect endangered species, and thus on to itself. The venture has been outstanding because those behind it have concentrated on selecting species which the Zoo can accommodate and manage best. Over two-thirds of the mammals, birds and reptiles which are now exhibited have been bred in the Zoo.

It was started at a time when it had become clear to all leading zoos that no single collection could hope to ensure the survival of any one endangered species. Out-breeding is essential if the genetic constitution of a rare species is not to be weakened, and this means that breeding groups need to be established in several zoos and that stock has to be exchanged. This fact is fully appreciated by the management of the Jersey Zoo. It is not an institution which hoards its surplus treasures.

The scale of the Zoo is in keeping with the size of the island. The 35 acres (14 hectares) of gently rolling countryside which the Zoo covers in the north-east corner of the island was once part of a property called Les Augres Manor. The old manor house, which is now the Zoo's administrative centre, dates from the latter half of the fourteenth century – the west wing of the original house still stands – and is one of the best examples of Jersey architecture. The closely-knit outbuildings are built of russet-coloured granite and have a warm and welcoming character, whatever the weather. They are tucked into the side of a gentle valley, through which flows a stream on its way to the sea a mile or so away. Nearly all of the Zoo's first buildings were either extensions or conversions of these outbuildings and of the walled garden of the manor house. Paddocks were then laid out along the same side of the valley, and the stream developed for wildfowl. New buildings were erected for spectacled bears, Brazilian tapirs and gorillas, and later still, on the opposite side of the stream from the manor house, a Marmoset House, a Reptile House, veterinary buildings and a small restaurant. Those original enclosures which have not proved adequate are now being replaced or rebuilt.

The Reptile House is Jersey's biggest building. Only half of the building is designed for the reptiles, the remainder being given over to food-preparation areas, offices and rooms for breeding. The old granite barn has been converted, with indoor dens and large outside cages for orang-utans on the ground floor, and the first floor housing such nocturnal mammals as hutias and tenrecs. Although the gorilla enclosure is only five years old, the breeding programme has already been so successful that the enclosure is now too small and new buildings are planned.

While it specializes in keeping endangered species, the Jersey Zoo is nevertheless a small general zoo that exhibits a variety of mammals, birds and reptiles. 35 of the approximately 125 species of animal which the Zoo holds are remnants of Gerald Durrell's original collection, and these are being gradually phased out to make room for species in greater danger. The remaining 90 are endangered species which need to be bred in captivity if they are to survive. Of the 90, 30 are mammals, 41 birds and 19 reptiles. 23 of the 30 mammalian species are breeding, 28 of the 41 birds, and 6 of the 19 reptile species, the most recent additions to the Zoo. So far the Jersey Wildlife Preservation Trust has been able to send to other zoos over 480 birds and mammals which have been bred on the island.

Jersey Zoo now has seven species of the two related groups of pocket-sized monkeys – marmosets and tamarins – from the forests of South America, and all but one of them breed regularly. Captive marmosets and tamarins need special care, and as yet only the common marmoset has been bred on a large scale, mainly for biological study. Both groups are becoming scarce in their natural habitat, but international agreement through the Washington Convention to Control Trade in Endangered Species has put a brake on their indiscriminate collection and export. Given scientific study and the right conditions, there is every reason to suppose that most species of tamarin and marmoset will breed as freely in captivity as the common marmoset already does.

The Jersey primate collection includes a group of Ursine black and white colobus monkeys from West Africa. This species is related to the more widely known East African black and white colobus, but it is becoming rare in its native habitat because the land is being developed for agricultural use. Colobus monkeys are difficult to keep because of their specialized leaf-eating habits. Like the Langur monkeys of Asia – and the cow – they have multi-chambered stomachs which harbour micro-organisms that help digestion, particularly of food with a high cellulose content that has to be broken down by microbial fermentation. The diet can only vary within narrow limits if the process of fermentation is not to be arrested, since the microbial population has to readjust every time the diet is changed. The group of colobus monkeys in Jersey consists today of six of the original group that was acquired and eleven animals that were born in the Zoo. The Zoo has also bred both the Bornean and Sumatran sub-species of orang-utan. For breeding purposes, the female Sumatran orang-utan belonging to Jersey is partnered by London Zoo's Gambar.

In 1972 four zoos in Europe, including Jersey and London, obtained groups of Mayotte brown lemurs, primitive members of the monkey family. These particular lemurs are indigenous to one of the Comoro Islands off the north-west coast of Madagascar, where their numbers have declined catastrophically, mainly because of trapping and shooting. The Jersey Mayotte lemurs have bred so successfully that they have trebled in number and Jersey, like London Zoo, has had to farm out specimens to other zoos to prevent overcrowding.

Three species of tenrecs are kept in the Jersey Zoo. These small creatures belong to the same order of mammals as hedgehogs and moles, but relatively little is known about them since they are rarely seen outside their native haunts in Madagascar and the Comoro Islands. Both the spiny tenrec and the pygmy tenrec have been successfully bred in the Zoo.

Another interesting animal kept in Jersey is the hutia, a rare rodent found only in the West Indies. Of ten species that have been described in scientific literature, only two are still relatively common. The others are listed as 'endangered'. Hutias are nocturnal and unprepossessing in appearance, and have never been well represented in zoos. All four of the species that are currently held in captivity in different zoos have bred. Jersey has both the Cuban and the Jamaican hutia, although only the latter has bred.

Jersey's pair of spectacled bears, a species which is fast becoming rare in its native haunts in South America, has raised four cubs. Another Jersey rarity is the Rodrigues fruit bat, an animal that is found only on Rodrigues Island, one of the Mascarene group between Mauritius and Reunion in the Indian Ocean. Today it is probably the rarest bat in the world, partly because it has been hunted, and partly because of the destruction of the forests which are its native habitat. On the other hand, the creature is not considered to have any unique zoological features, and it has a number of close and common relatives that are classified in the same genus. Jersey has the only breeding group in captivity in the world.

Most of the birds which are kept in the Zoo are rare in the wild. They include waldrapp, Congo peafowl, pink pigeon, Rodrigues' fody, St Lucia and St Vincent Amazon parrots and white eared pheasants. The most conspicuous breeding success has been that of the white eared pheasant, a native of China. In 1966, when Jersey acquired two pairs that had been hatched in the Peking Zoo, there were thought to be only eighteen specimens outside China. One of the males sent to Jersey died almost immediately, and a female a year later. When the cock of the surviving pair damaged a leg, all seemed to be lost. But in 1969 a clutch of nineteen eggs was unexpectedly laid, of which sixteen hatched and thirteen birds survived. Since then over a hundred of the pheasants have been successfully reared. Of these seventy-five have been sent to other collections.

The Jersey Zoo has instituted a useful system of data recording for all its animals. The information it provides is collated in articles that are published, mostly in the Zoo's *Annual Report*.

What is so encouraging about the Jersey Wildlife Preservation Trust is that it has not set itself up in isolation. From the start it has co-operated with other well-managed collections from whom, in return, it has received much technical help. The respect in which the Trust is held was marked in 1972 when Jersey was chosen as the venue of a conference on the breeding of endangered species. Various governments have also shown their faith in the Jersey venture by placing animals 'on loan' with the Trust. For example, breeding stocks of pink pigeon, Guenther's gecko, Telfair's skink and Rodrigues fruit bat have been lent by the government of Mauritius, and seven St Lucia parrots have been lent by that island's administration.

Jurong Bird Park, Singapore

Peter Olney

Jurong Bird Park is a good example of a collection confined to one order of animals. Over 7,000 birds from approximately 350 species and subspecies are kept in 78 aviaries dotted around the Park. Not surprisingly, South-east Asian species predominate, and there are particularly interesting and valuable groups of birds-of-paradise, hornbills, parrots and pheasants. The Park covers 50.3 acres (20.3 hectares) on the western slopes of Jurong Hill, some fifteen miles (24 kilometres) from the centre of the city of Singapore. Built at a cost of five million Singapore dollars, the Park took two years to construct, and was officially opened to the public on 3 January 1971. It is run by an entirely government-owned company, which in the main relies for its finances on gate revenue. In the first seven years after it was opened, over two million people visited the Park.

The Park has been well designed and landscaped, and most of the aviaries merge into a background of trees and shrubs. A road winds past the aviaries, and there is a trackless tram-car which runs on a circular tour. From a restaurant on the edge of the largest lake, which contains swans, pelicans and flamingos, there are superb views of the terraced slopes. An area which is closed to the public has been set aside for nesting birds and for those in transit or in quarantine.

The Park is known not only for its comprehensive bird collection, but also for a remarkable engineering and architectural feat – the spectacular Free-Flight Aviary which covers some 5 acres (2 hectares) and which contains what is probably the highest man-made waterfall in the world. To support the huge mesh that covers the Aviary there are eighteen steel cables each weighing about three tons. The waterfall, which dominates the Aviary and the Park, has a solid backdrop of boulders held together by steel cables. Filtered water cascades over a 100-foot (30.5-metre) drop to the rocks below before flowing into a series of pools and streams. A visitor within the Aviary does not have any impression that he is in a man-made structure. Oriental bridges take visitors over the streams among the lush vegetation. There are said to be over 3,000 birds of about 100 species in this aviary alone, but an accurate census in such a large area would be impossible. Gregarious birds such as the beautiful green parakeets and white egrets, are among the more conspicious inhabitants.

Jurong Park, although one of the newest collections in the world, could provide enviable opportunities for serious scientific research, and the company hopes that funds will soon be forthcoming from the government for this purpose.

Parc Zoologique de Clères

Peter Olney

Another internationally famous bird collection, that at Clères in northern France, was started in 1919 by Dr Jean Delacour, the doyen of aviculturists. While he still owns the property, which has been his home since 1920, he has bequeathed the collection to the Muséum National d'Histoire Naturelle in Paris. The château and its park, which have been described by Colette as 'Paradis Terrestre', form a perfect setting for the collection of animals. Half of the 65 acres (26 hectares) of parkland is open to the public while the other half constitutes an undisturbed breeding area.

The château, the greater part of which dates from the thirteenth and fourteenth centuries, is surrounded by trees and grassy slopes which stretch down to a lake and a stream. The whole area has been so well designed that château, plants, animals, aviaries and paddocks blend into the landscape. The château only just survived the Second World War; there was a disastrous fire in 1939, and it was bombed in 1940. However, due to Jean Delacour's energy and dedication, by 1947 the collection had been re-established. In order to help finance the Park, what was once a private collection is now open to the public. In 1976 over 145,000 people paid to visit Clères.

Before the war the collection was probably the finest of its kind in the world, containing as it did many rarely-seen species brought back by Jean Delacour from his extensive travels. Some species were completely new to science, such as the two imperial pheasants which he brought back from central Vietnam in 1924, and which have not been seen there, or anywhere else, since. The descendants of these two birds can now be seen in collections in other parts of the world.

Today, when national and international import restrictions make it difficult to obtain birds, the Clères collection is still one of the best in the world. There are some 2,000 individuals of 400 species, mainly waterfowl and pheasants, but there are also cranes, flamingos, pigeons and doves, turacos and parrots. A high proportion of the birds breed regularly, including such rarities as Cuban whistling duck, Hawaiian geese, red-breasted geese, white-winged wood duck, Edward's pheasant, and Palawan peacock pheasant.

Among the few mammals kept in the Park, either in large enclosures or roaming freely, are Bennett's wallabies, axis deer, blackbuck and sitatunga.

The collection has always been used for research, especially on behaviour, on species affinities, and on breeding. Jean Delacour and his collaborators have contributed many articles to scientific journals over the years. He was editor of *L'Oiseau* from 1920 to 1940, and has written many books, including *The Waterfowl of the World*, in four volumes, *Les Oiseaux de l'Indochine Française*, *The Birds of Malaya* and *The Pheasants of the World*.

The Wildfowl Trust, Slimbridge, Gloucestershire

Peter Olney

Perhaps the best-known example of a collection devoted to a single group of animals is the Wildfowl Trust, formed by Sir Peter Scott in 1946. The objects of the Trust are to maintain and breed wildfowl, especially endangered species, to carry out scientific research, to advise on wildfowl conservation, and to educate the public to a greater appreciation of nature in general, and of wildfowl in particular.

Wildfowl comprise ducks, geese and swans, and all belong to one family, the *Anatidae*. This family, which contains about 140 different species with about 250 different types (subspecies or races) is relatively discrete in that all members have many characters in common, although they may look and behave differently. Affinities within the family are relatively easy to recognize, and because of the close relationship of many species, comparative work on their morphology, ecology and behaviour is particularly valuable to taxonomists and behaviourists.

Wildfowl are found throughout the world, but usually in areas that are especially prone to interference by man. Fortunately, most wildfowl, although often shy and wary in the wild, adapt easily to captivity and, given the right conditions, will breed well.

The Trust's headquarters were set up on low-lying ground on the banks of the River Severn near Slimbridge in Gloucestershire. For centuries this area has been a natural feeding and resting ground for birds, especially wintering and migrating wildfowl, and for flocks of white-fronted geese, wigeon, teal and other ducks.

The ground was laid out in a series of inter-communicating pools and streams which run through open and enclosed areas planted with appropriate marshland vegetation. The collection now numbers nearly 3,000 birds of some 200 species and subspecies. Although by far the majority are wildfowl, there are also representatives of their close relatives, the screamers of South America, and the six types of flamingo that are known to exist. In 1967 a heated Tropical House was opened for exotic ducks and a variety of small birds such as hummingbirds, tanagers, sunbirds and doves.

The Trust's record of breeding successes is outstanding. All 8 species of swan, 4 types of flamingo, and all but a few of the 132 geese and duck species have bred, many of them for the first time in captivity. Possibly the most spectacular success has been with the Hawaiian goose. Confined to Hawaii, it was on the verge of extinction in 1952, with only an estimated 30 left in the world. The few imported to Slimbridge bred so well that the Trust was not only able to supply many other collections, but also returned 200 of the birds to Hawaii in an effort to re-establish them in their natural habitat.

Most of the wild geese and ducks feed in the area of marshland and pasture which is outside the enclosures and, in order not to disturb the birds, the visitor is led to carefully positioned hides by routes screened by earthbanks. it is estimated that over 40,000 people go to see the wild geese, swans and ducks during the six months of the autumn and winter.

Although Slimbridge was the first of the Wildfowl Trust's collections, it has several offspring. Martin Mere, in Lancashire, which is a natural wintering ground, particularly for pink-footed geese, has been developed along similar lines to Slimbridge. At Arundel in Sussex, at Peakirk in Cambridgeshire and at Washington in Tyne and Wear the centres have placed more emphasis on keeping tame birds. Apart from the general educational value of the collections, each of the five centres provides a formal education programme and has its own staff, who give lectures and demonstrations to schoolchildren of all ages and to adult education groups.

There are also two 'refuges' owned or controlled by the Trust: 700 acres (280 hectares) at Welney in Cambridgeshire, and 600 acres (240 hectares) at Caerlaverock in Dumfriesshire. These areas are crucial to the preservation of wildfowl, giving protection to many thousands of birds including, at Caerlaverock, most of the barnacle geese from Spitzbergen. The Trust also controls bird-ringing stations at Borough Fen in Cambridgeshire, Abberton Reservoir in Essex and Nacton in Suffolk.

The Trust has always regarded research as one of the most important aspects of its work, and studies have been carried out on population dynamics, foods, navigation, migration, behaviour and breeding biology. A new research block at Slimbridge contains laboratories, a library and a museum, offices and lecture rooms. Research results are published in a number of scientific journals, as well as in *Wildfowl*, the Trust's own annual report. The major part of the Trust's finances come from public entrance fees (nearly half a million people visited the various centres in 1977), and from the subscriptions paid by its 16,000 members.

The stability of the Trust, and its international standing as a research and conservation centre, owes much to Sir Peter Scott's enthusiasm and dedication. 'Ours [the zoological collections] serve their purpose if they arouse the interest of public and politicians and ensure the conservation of the wild wetlands on which the waterfowl depend,' he wrote, in an article entitled 'Reconciling the Irreconcilable – Wildfowl and People': a task to which the Trust and its staff have been dedicated for over thirty years.

Arizona-Sonora Desert Museum, Tucson, Arizona

Peter Olney

The great Sonora Desert, which straddles the south-western border that divides the United States from Mexico, is the setting for one of the most interesting specialist zoos. Fifteen miles (24 kilometres) west of Tucson, in Tucson Mountain Park, lies a magnificent expanse of arboreal desert: an area of red and purple ridges, covered by a forest of tall green saguaro cacti. In order to survive, plants and animals have adapted themselves in various ingenious ways to this region where rainfall averages less than 10 inches (25.4 centimetres) a year.

William H. Carr, the man who first thought of establishing a museum in the desert, had been responsible, on behalf of the American Museum of Natural History, for the development of the Trailside Museum in Bear Mountain State Park, 40 miles (64 kilometres) north of New York City. This was one of the first displays in the world to combine natural trails, animal exhibits and a museum, in order to demonstrate local wildlife. Carr's idea was that a similar museum should be set up to display the unique and highly individual existence of the animals and vegetation of the Sonora Desert, and in 1951 he put his idea to the Pima County Park Advisory Committee. One of the members of this Committee was Arthur N. Pack, the founder of the American Nature Association who, on behalf of the Association, offered an initial $10,000 so that work on the Museum could be started. Other contributions, totalling $450,000, were raised through the Association and six months later, on Labor Day 1962, the Museum was opened, with Carr as its Director and Pack as its President.

The interest and enthusiasm shown by the public from the very beginning surprised everyone concerned with the enterprise. By 1977 six million people had visited the Museum. With fees coming in from members throughout the world, and numerous bequests, which together equalled the money taken at the gate, the Museum was financially sound. Its governing board was even thinking of limiting the number of visitors to 400,000 or so a year, an almost unheard-of idea in zoo policies.

Representatives of about 200 species of animals – from invertebrates and fish of the Gulf of California to Desert bighorn sheep – all of which are native to the States of Arizona in the USA and Sonora in Mexico, are exhibited in conditions which essentially reproduce the habitats found in the region, zoologically one of the richest in America. As well as desert areas, there is subtropical swampland, deciduous woodland, prairie, cold-water streams and pools, alpine meadows and crags. For example, in the recently completed glassed-in grottoes for small cats, ocelots, bobcats, margays and jaquarundis live in surroundings that look so natural that visitors have been convinced that everything is real. In fact, only the cats and the native flora are real; the rest – the massive outcroppings of rocks and boulders which cover many thousands of square feet – is made of coloured concrete on steel frames. The various tableaux have also been designed with great care and accuracy, and display such subjects as lac insects and carpenter ants.

Near the tableaux is an area, designed to simulate the banks of the Santa Cruz River, where otters and beavers live in pools shaded by cottonwood trees. The animals can be seen under water and in their dens, and the beautiful setting makes this one of the finest aquatic mammal exhibits in the world.

Another very popular exhibit, 'the tunnel in the desert', came about as a direct result of visitors wanting to see nocturnal and burrowing animals. A long gallery was excavated 12 feet (3.7 metres) underground. One side is lined with faithfully reproduced caves and burrows, tree hollows and dens, and the other by roots of living trees grown in pits opening to the outside surface. Each display has its individual lighting system, which is worked by the visitor, and a label which explains the ways in which vegetation and animals have adapted to desert life. Among the creatures shown in the tunnel are skunks, cacomistles, badgers, prairie dogs, porcupines, rattlesnakes, ants and bees, and many of them, including the increasingly rare kit fox, the fruit bats and vampire bats, have successfully bred.

The Gulf of California and its islands, which comprise sea surrounded almost entirely by desert, is one of the latest areas to be interpreted by the Museum. A horn shark and a green turtle are two recent acquisitions, and the tidepool display is filled with small fish, crabs, anemones, starfish and sea-urchins from the shores of Puerto Penasco.

The Earth Sciences Department, with its Congdon Earth Sciences

Center, is a logical extension of the Museum's other interests. Applying the same methods of naturalistic simulation coupled with explanatory models, artwork and texts, the Center demonstrates the dynamics of the earth's formation and the erosive forces that have shaped the desert.

A zoo such as the Desert Museum, which exhibits only the local fauna, can be a potent force for education and conservation. One of the practical ways in which the Museum helps is by answering queries from people in the area who want to know how to deal with such creatures as rattlesnakes or tarantulas. The Demonstration Desert Garden, built in 1963, shows how shrubs and trees native to Arizona and Sonora can be adapted for domestic landscaping. In this way the Museum can also educate and influence the people who hold in their hands, more so than most, the fate of the animals and plants that comprise their immediate environment. Apart from the knowledge that visitors gain from seeing living illustrations of the patterns of inter-relationships between animals and nature, the Museum has a formal programme of lectures and tours. A small fleet of mobile classrooms goes from school to school, both in Arizona and over the border into Mexico. Almost from the Museum's inception, there has been a strong connection between Sonora and Arizona, and the flags of both nations and both states fly at the entrance gates. In addition to the bilingual mobile classes, there are reciprocal study visits, research projects, collecting trips, information exchanges and a continuing series of bi-national conferences on the threatened environment of the Gulf of California.

Among the Museum's recent and current research activities are the study of the biology of the margay in Sonora, the nesting behaviour of red-tailed hawks and Harris's hawks, the use of mesquite, seagrasses and temparies by south-west Indian cultures, the ethno-botany of the Serie Indians, and the method of growing nutritive food crops in waterless regions.

Like most zoos, specialist collections often suffer financial problems. Those which have to rely entirely on gate revenue have to make the exhibits interesting enough to attract sufficient visitors. Some are fortunate enough to be subsidized either in whole or in part by individuals (often the owners); by membership fees; by grants from research bodies, education authorities, charitable organizations, or municipal bodies. For example, the International Crane Foundation in Wisconsin, which carries out research on the biology and conservation of cranes, is a private, non-profit making organization which relies on donations, membership fees and grants. The Mount Hagen Sanctuary in New Guinea, which specializes in birds-of-paradise and indigenous fauna, is a government-appointed and government-aided Trust. Twycross Zoo, in England, which began as a private collection, has had to expand and bring in more exotic animals in order to attract more visitors. And the Pheasant Trust, also in England, continues to exist in an established general collection only because of the generosity and enthusiasm of the founder and his fellow members.

Introduction: References

1 CHALMERS MITCHELL, P., 1929, *Centenary History of the Zoological Society of London*, London.

2 COLE, F. J., 1944, *A History of Comparative Anatomy*, London.

3 COLINVAUX, P., 1978, *Why Big Fierce Animals Are Rare*, New Brunswick.

4 CRANDALL, L. S., 1964, *The Management of Wild Mammals in Captivity*, Chicago and London.

5 DE BEER, G., 1955, *Alps and Elephants: Hannibal's March*, London.

6 DELEUZE, J. P. F., 1823, *History and Description of the Royal Museum of Natural History*, Paris.

7 ENCYCLOPAEDIA BRITANNICA, 11th and 15th editions.

8 GEOFFROY-SAINT-HILAIRE, ÉTIENNE, 1833, *Proc. Zool. Soc.*, p. 15.

9 HARCOURT, A., and K. STEWART, 1978, 'By courtesy of man', *New Scientist* (9 March), 644–5.

10 HEDIGER, H., 1969, *Man and Animal in the Zoo*, trans. G. Vevers and W. Reade, London.

11 JORDAN, B., and S. ORMROD, 1978, *The Last Great Wild Beast Show*, London.

12 KIRCHSHOFFER, R., (ed.), 1968, *The World of Zoos*, Intro. by Dr B. Grzimek, trans. H. Morris, London.

13 KLINGENDER, F., 1971, *Animals in Art and Thought*, ed. E. Antal and J. Harthan, London.

14 *International Zoo Yearbook*, 1960–1977, Vols 1–17, London.

15 LEMOINE, P., 1935, 'Le Muséum National d'Histoire Naturelle', *Arch. mus. natn. hist. nat., Paris* (6) *12*, (1), pp. 1–79.

16 LOISEL, G., 1912, *Histoire des Ménageries de l'Antiquité à nos Jours*, 1, Paris.

17 MONTAGU, M. F. A., 1943, 'Edward Tyson 1650–1708 and the rise of human and comparative anatomy in England', *Mem. Amer. Philos. Soc.*, 20.

18 NORDENSKIÖLD, E., 1929, *The History of Biology*, New York and London.

19 RACKHAM, H. (trans.), 1940, *Pliny. Natural History*, 3. London and Cambridge, Mass.

20 *Reports of the Zoological Society for the Year 1929*, 1930, London.

21 ROSE, K., 1978, *William Harvey*, London.

22 SILLAR, F. C., and R. M. MEYLER, 1968, *Elephants Ancient and Modern*, London.

23 SMITH, A., and J. SOUTHAM, 1977, *Animals on View*, London.

24 TOYNBEE, J. M. C., 1973, *Animals in Roman Life and Art*, London.

25 WEMMER, C., 1977, 'Can wildlife be saved in zoos?', *New Scientist* (8 September), 585–7.

Zoo Chapters: Bibliography

The main sources are listed. A considerable amount of information was also taken from zoo journals and magazines, guidebooks, annual and scientific reports, Council minutes, progress reports and feasibility studies. Much help, often in the form of unpublished material, was also given by zoo personnel.

ANON, 1976, *Zoo Book*, Smithsonian Institution, Washington.

BEEBE, W., 1918–22, *A Monograph of the Pheasants*, Vols 1–4, London.

BENNETT, E. T., 1830–1, *The Gardens and Menagerie of the Zoological Society Delineated*, 2 vols, London.

BIGALKE, R., 1954, *The National Zoological Gardens of South Africa*, Pretoria.

BOITARD, P., 1942, *Le Jardin des Plantes*, Paris.

BLUNT, W., 1976, *The Ark in the Park*, London.

CHALMERS MITCHELL, P., 1929, *Centenary History of the Zoological Society of London*, London.

CRANDALL, L. S., 1964, *The Management of Wild Animals in Captivity*, Chicago and London.

DELACOUR, J., 1966, *The Living Air*, London.

DELEUZE, J. P. F., 1823, *History and Description of the Royal Museum of Natural History*, Paris.

DITTRICH, L., 1966, *Auf Safari in Europa*, Hanover.

FISHER, J., 1966, *Zoos of the World*, London.

GEIGY, R., E. M. LANG, H. WACKERNAGEL, P. STUDER and K. BRAGGER, 1974, *100 Jahre Zoologischer Garten Basel*, Basle.

HAHN, E., 1968, *Zoos*, London.

HOLLOWAY, S., 1976, *The London Zoo*, London.

INTERNATIONAL ZOO YEARBOOK, 1960–79, Vols 1–19, London.

KELLER, R., and C. R. SCHMIDT, 1978, *Das Buch vom Zoo*, Frankfurt.

KIRCHSHOFFER, R. (ed.), 1968, *The World of Zoos*, London.

KLOS, H.-G., 1969, *Von der Menagerie zum Tierparadies*, Berlin.

LEMOINE, P., 1935, 'Le Muséum National d'Histoire Naturelle. Son Histoire – Son État Actuel', *Archs Mus. natn. Hist. nat., Paris* (6) 12 : 1–79.

LOISEL, G., 1912, *Histoire des ménageries de l'antiquité à nos jours*, 3 vols, Paris.

MILNE-EDWARDS, A., E. BLANCHARD and G. P. DESHAYES, 1871, 'Rapport sur les dégâts occasionés dans le département zoologique du Muséum par le bombardement de cet établissement scientifique par les Prussiens', *Nouv. Archs Mus. Hist. nat., Paris 6* 1870 (Bull. No. 6 : 29–32).

BIBLIOGRAPHY

PEEL, C. V. A., 1903, *The Zoological Gardens of Europe*, London.

ROUSSEAU, L., and C. LEMONNIER, 1837, *Promenades au Jardin des Plantes, comprenant la description de la ménagerie, etc.*, Paris.

SCHERREN, H., 1905, *The Zoological Society of London*, London.

SCLATER, P. L., (ed.), 1901, *A Record of the Progress of the Zoological Society of London during the Nineteenth Century*, London.

STREET, P., 1953, *Whipsnade*, London.

URBAIN, A., 1935, 'Leçon inaugurale du Cours d'Ethologie des animaux sauvages', *Archs Mus. natn. Hist. nat., Paris* (6) 12 : 1–79.

VEVERS, G., 1976, *London's Zoo*, London.

WEGEFORTH, H. M., and N. MORGAN, 1969, *It Began with a Roar*, San Diego.

ZUCKERMAN, S. (ed.), 1976, *The Zoological Society of London 1826–1976 and Beyond*, London.

Scientific Names of Animals Where Species or Genus Identified

FISH

CHINESE STURGEON	*Acipenser* sp.
GOLDFISH	*Carassius auratus*

AMPHIBIA

CHINESE GIANT SALAMANDER	*Andrias davidianus*
JAPANESE GIANT SALAMANDER	*Andrias japonicus*

REPTILES

ALLIGATOR	
Mississippi alligator	*Alligator mississippiensis*
Yangtse (Chinese) alligator	*Alligator sinensis*
CROCODILE	
African dwarf crocodile	*Osteolaemus tetraspis*
American crocodile	*Crocodylus acutus*
GECKO	
Guenther's gecko	*Phelsuma guentheri*
Tokay gecko	*Gekko gecko*
GHARIAL	*Gavialis gangeticus*
KOMODO DRAGON	*Varanus komodoensis*
LIZARD	
Frilled lizard	*Chlamydosaurus kingi*
Monitor	*Varanus* sp.
Nile monitor	*Varanus niloticus*
SNAKE	
Boa constrictor	*Boa constrictor*
Mamba	*Dendroaspis* sp.
Mangrove snake	*Boiga dendrophila*

Taipan	*Oxyuranus scutellatus*
Tiger snake	*Notechis scutatus*
TELFAIR'S SKINK	*Leiolopisma telfairii*
TORTOISE	
Aldabra tortoise	*Geochelone gigantea*
Galapagos tortoise	*Geochelone elephantopus*
TUATARA	*Sphenodon punctatus*
TURTLE	
Alligator snapping turtle	*Macrochelys temmincki*
Green turtle	*Chelonia mydas*

BIRDS

BIRD OF PARADISE	
Lesser bird of paradise	*Paradisaea minor*
BOWERBIRD	
Satin bowerbird	*Ptilonorhynchus voilaceus*
BUSTARD	
Great bustard	*Otis tarda*
BUZZARD	
Long-legged buzzard	*Buteo rufinus*
CASSOWARY	*Casuarius* sp.
COCKATOO	*Cacatua* sp.
CONDOR	
Andean condor	*Vultur gryphus*
CORMORANT	*Phalacrocorax* sp.
CRANE	
Crowned crane	*Balearica* sp.
Demoiselle crane	*Anthropoides virgo*
Japanese (Manchurian) crane	*Grus japonensis*
Sarus crane	*Grus antigone*
White-naped crane	*Grus vipio*
DODO	*Raphus cucullatus*
DUCK	
Cuban whistling duck	*Dendocygna arborea*
White-winged wood duck	*Cairina scutulata*

SCIENTIFIC NAMES OF ANIMALS

EAGLE	
Bald eagle	*Haliaeetus leucocephalus*
Monkey-eating eagle	*Pithecophaga jefferyi*
Tawny eagle	*Aquila rapax*
EGRET	
Cattle egret	*Ardeola ibis*
White egret	*Egretta alba*
EMU	*Dromais novaehollandiae*
FLAMINGO	
Chilean flamingo	*Phoenicopterus chilensis*
Greater flamingo	*Phoenicopterus ruber roseus*
Rosy flamingo	*Phoenicopterus ruber ruber*
FODY	
Rodrigues' fody	*Foudia flavicans*
GANNET	*Morus bassanus*
GOOSE	
Barnacle goose	*Branta leucopsis*
Cereopsis goose	*Cereopsis novaehollandiae*
Hawaiian goose	*Branta sandvicensis*
Pink-footed goose	*Anser fabalis brachyrhynchus*
Red-breasted goose	*Branta ruficollis*
White-fronted goose	*Anser albifrons*
HAWK	
Harris's hawk	*Parabuteo unicinctus*
Red-tailed hawk	*Buteo jamaicensis*
HERON	
Night heron	*Nycticorax nycticorax*
HOATZIN	*Opisthocomus hoazin*
HONEYEATER	*Meliphaga* sp.
HORNBILL	
Southern pied hornbill	*Anthracoceros coronatus convexus*
Yellow-billed hornbill	*Tockus flavirostris*
IBIS	
Red-cheeked ibis (Waldrapp)	*Geronticus eremita*
Sacred ibis	*Threskiornis aethiopicus*
Straw-necked ibis	*Carphibis spinicollis*
KEA	*Nestor notabilis*
KIWI	*Apteryx australis*

LYREBIRD	
Superb lyrebird	*Menura novaehollandiae*
MALLEE FOWL	*Leipoa ocellata*
NIGHTJAR	
Owlet nightjar	*Aegotheles cristatus*
OSTRICH	*Struthio camelus*
OWL	
Snowy owl	*Nyctea scandiaca*
PARROT	
Golden-shouldered parrot	*Psephotus c. chrysopterygius*
Hooded parrot	*Psephotus c. dissimilis*
Mulga parrot	*Psephotus varius*
Paradise parrot	*Psephotus pulcherrimus*
St Lucia Amazon parrot	*Amazona versicolor*
St Vincent Amazon parrot	*Amazona guildingii*
PEAFOWL	
Congo peafowl	*Afropavo congensis*
PELICAN	
American white pelican	*Pelecanus erythrorhynchos*
Brown pelican	*Pelecanus occidentalis*
Dalmatian pelican	*Pelecanus crispus*
Philippine pelican	*Pelecanus philippensis*
Pink-backed pelican	*Pelecanus rufescens*
White pelican	*Pelecanus onocrotalus*
PENGUIN	
Adelie penguin	*Pygoscelis adeliae*
Black-footed (Jackass) penguin	*Spheniscus demersus*
Humboldt's penguin	*Spheniscus humboldti*
King penguin	*Aptenodytes patagonica*
PHEASANT	
Blue eared pheasant	*Crossoptilon auritum*
Brown eared pheasant	*Crossoptilon mantchuricum*
Cheer pheasant	*Catreus wallichi*
Edwards' pheasant	*Lophura edwardsi*
Elliot's pheasant	*Syrmaticus ellioti*
Imperial pheasant	*Lophura imperialis*
Mikado pheasant	*Syrmaticus mikado*
Palawan peacock pheasant	*Polyplectron emphanum*
Silver pheasant	*Lophura nycthemera*
White eared pheasant	*Crossoptilon crossoptilon*

PIGEON	
Pink pigeon	*Columba mayeri*
RAVEN	*Corvus corax*
RHEA	*Rhea americana*
ROADRUNNER	*Geococcyx californiana*
SCRUB-WREN	*Sericornis* sp.
SPOONBILL	*Platalea leucorodia*
SWAN	
Black swan	*Cygnus atratus*
Trumpeter swan	*Cygnus cygnus buccinator*
TAWNY FROGMOUTH	*Podargus strigoides*
TERN	
Inca tern	*Larosterna inca*
THORNBILL	*Acanthiza* sp.
TURACO	*Tauraco* sp.
VULTURE	
Black vulture	*Coragyps atratus*
Griffon vulture	*Gyps fulvus*
King vulture	*Sarcoramphus papa*
Turkey vulture	*Cathartes aura*
WALDRAPP (Red-cheeked ibis)	*Geronticus eremita*

MAMMALS

AARDVARK	*Orycteropus afer*
AARDWOLF	*Proteles cristatus*
ADDAX	*Addax nasomaculatus*
ALPACA	*Lama pacos*
ANOA	
Mountain anoa	*Anoa quarlesi*
ANTEATER	
Giant anteater	*Myrmecophaga tridactyla*
ANTELOPE	
Nilgai (Blue) antelope	*Boselaphus tragocamelus*
Nyala	*Tragelaphus angasi*
Pronghorn antelope	*Antilocapra americana*

SCIENTIFIC NAMES OF ANIMALS

ARMADILLO	
Nine-banded armadillo	*Dasypus novemcinctus*
ASS	
Asiatic wild ass	*Equus hemionus*
Somali wild ass	*Equus africanus somaliensis*
AUROCH	*Bos primigenius*
AYE-AYE	*Daubentonia madagascariensis*
BANTENG	*Bos javanicus*
BARASINGHA (Swamp deer)	*Cervus duvauceli*
BAT	
Rodrigues' fruit bat (Flying fox)	*Pteropus rodricensis*
Vampire bat	*Desmodus rotundus*
BEAR	
Asiatic black bear	*Selenarctos thibetanus*
Caucasus brown bear	*Ursus arctos caucasicus*
Cinnamon brown bear	*Ursus arctos cinnamomum*
European brown bear	*Ursus arctos arctos*
Grizzly bear	*Ursus arctos horribilis*
Kodiak bear	*Ursus arctos middendorffi*
Kamchatka bear	*Ursus arctos beringianus*
American black bear	*Euarctos americanus*
Polar bear	*Thalarctos maritimus*
Spectacled bear	*Tremarctos ornatus*
Sun bear	*Helarctos malayanus*
BEAVER	*Castor* sp.
BETTONG (Brush-tailed rat-kangaroo)	*Bettongia penicillata*
BHARAL (Blue sheep)	*Pseudois nayaur*
BISON	
American bison	*Bison bison*
European bison (Wisent)	*Bison bonasus*
Wood bison	*Bison bison athabascae*
BLACKBUCK	*Antilope cervicapra*
BLESBOK	*Damaliscus dorcas phillipsi*
BOBCAT	*Felis rufa*
BONGO	*Boocercus eurycerus*
BONTEBOK	*Damaliscus dorcas dorcas*

BUFFALO	
Cape buffalo	*Syncerus caffer caffer*
Congo buffalo	*Syncerus caffer nanus*
BUSHBABY	*Galago* sp.
CACOMISTLE	*Bassariscus astutus*
CAMEL	
Arabian camel	*Camelus dromedarius*
CARIBOU (Reindeer)	*Rangifer tarandus*
CAT	
Geoffroy's cat	*Felis geoffroyi*
Leopard cat	*Felis bengalensis*
Serval cat	*Felis serval*
CHAMOIS	*Rupicapra rupicapra*
CHEETAH	*Acinonyx jubatus*
CHIMPANZEE	
Chimpanzee	*Pan troglodytes*
Pygmy chimpanzee	*Pan paniscus*
COYOTE	*Canis latrans*
DEER	
Axis deer	*Axis axis*
Barking deer	*Muntiacus* sp.
Brow-antlered (Eld's) deer	*Cervus eldi*
Fallow deer	*Dama dama*
Hog deer	*Axis porcinus*
Indian swamp deer (Barasingha)	*Cervus duvauceli*
Indo-China Sika deer	*Cervus nippon pseudaxis*
Mule deer	*Odocoileus hemionus*
Musk deer	*Moschus moschiferus*
Père David's deer	*Elaphurus davidianus*
Red deer (Wapiti)	*Cervus elaphus*
Red brocket deer	*Mazama americana*
River (Chinese water) deer	*Hydropotes inermis*
Roe deer	*Capreolus capreolus*
White-lipped deer	*Cervus albirostris*
White-tailed deer	*Odocoileus virginianus*
DIK-DIK	*Madoqua* sp.
DINGO	*Canis familiaris dingo*
DOG	
African (Cape) wild dog	*Lycaon pictus*

DOLPHIN	
Bottle-nosed dolphin	*Tursiops truncatus*
Chinese river dolphin	*Lipotes vexillifer*
DRILL	*Mandrillus leucophaeus*
DUIKER	
Common (Grey) duiker	*Sylvicapra grimmia*
ECHIDNA	
Australian (Common) echidna	*Tachyglossus aculeatus*
ELAND	*Taurotagus oryx*
ELEPHANT	
African elephant	*Loxodonta africana*
Indian elephant	*Elephas maximus*
ELK	
American elk	*Cervus elaphus canadensis*
FANALOKA	*Fossa fossa*
FOSSA	*Cryptoprocta ferox*
FOX	
Fennec fox	*Fennecus zerda*
Kit fox	*Vulpes velox*
GAUR	*Bos gaurus*
GAZELLE	
Grant's gazelle	*Gazella granti*
Slender-horned gazelle	*Gazella leptoceros*
Speke's gazelle	*Gazella spekei*
Thomson's gazelle	*Gazella thomsoni*
GEMSBOK (Cape oryx)	*Oryx gazella gazella*
GERENUK	*Litocranius walleri*
GIANT PANDA	*Ailuropoda melanoleuca*
GIRAFFE	*Giraffa camelopardalis*
GNU	
Brindled gnu	*Connochaetes taurinus*
White-tailed gnu	*Connochaetes gnou*
GOAT	
Cretan goat	*Capra aegagrus cretica*
Rocky mountain goat	*Oreamnos americanus*
GORAL	*Nemorhaedus goral*

GORILLA	
Lowland gorilla	*Gorilla gorilla gorilla*
GUANACO	*Lama guanicoe*
HARTEBEEST	
Cape hartebeest	*Alcelaphus caama*
Lichtenstein's hartebeest	*Alcelaphus lichtensteini*
HIPPOPOTAMUS	
Hippopotamus	*Hippopotamus amphibius*
Pygmy hippopotamus	*Choeropsis liberiensis*
HUTIA	
Cuban hutia	*Capromys pilorides*
Jamaican hutia	*Geocapromys browni*
IBEX	*Capra ibex*
ICHNEUMON (Egyptian mongoose)	*Herpestes ichneumon*
IMPALA	*Aepyceros melampus*
JACKAL	
Black-backed jackal	*Canis mesomelas*
JAGUAR	*Panthera onca*
JAQUARUNDI	*Felis yagouaroundi*
KIANG (Tibetian wild ass)	*Equus hemionus kiang*
KLIPSPRINGER	*Oreotragus oreotragus*
KOALA	*Phascolarctos cinereus*
KUDU	*Tragelaphus* sp.
KULAN (Turkmenian wild ass)	*Equus hemionus kulan*
LANGUR	
Douc langur	*Pygathrix nemaeus*
Golden langur	*Presbytis gaei*
White-headed langur	*Trachypithecus leucocephalus*
LECHWE	*Kobus* sp.
LEMUR	
Mayotte brown lemur	*Lemur macaco mayottensis*
LEOPARD	
Black leopard	*Panthera pardus*
Chinese leopard	*Panthera pardus japonensis*
Clouded leopard	*Neofelis nebulosa*
Snow leopard	*Panthera uncia*

LION	
Lion	*Panthera leo*
Indian lion	*Panthera leo persica*
LLAMA	*Lama glama*
LYNX	
Caracal lynx	*Felis caracal*
MACAQUE	
Japanese macaque	*Macaca fuscata*
MANATEE	
Caribbean manatee	*Trichechus manatus*
MANDRILL	*Mandrillus sphinx*
MANGABEY	*Cercocebus* sp.
MARGAY	*Felis wiedi*
MARKHOR	*Capra falconeri*
MINK	*Mustela lutreola*
MOLE	
Marsupial mole	*Notoryctes typhlops*
MONGOOSE	
Yellow mongoose	*Cynictis penicillata*
MONKEY	
De Brazza's monkey	*Cercopithecus neglectus*
Proboscis monkey	*Nasalis larvatus*
Snub-nosed monkey	*Rhinopithecus roxellanae*
Tcheli monkey	*Macaca mulatta*
Ursine black & white colobus monkey (King colobus)	*Colobus polykomos polykomos*
Vervet monkey	*Cercopithecus aethiops*
MUNTJAC	*Muntiacus* sp.
MUSKRAT	*Ondatra zibethica*
NUMBAT	*Myrmecobius fasciatus*
OCELOT	*Felis pardalis*
OKAPI	*Okapia johnstoni*
ONAGER	*Equus hemionus onager*
ORANG-UTAN	
Bornean orang-utan	*Pongo pygmaeus pygmaeus*
Sumatran orang-utan	*Pongo pygmaeus abeli*

ORYX	
Arabian oryx	*Oryx leucoryx*
Cape oryx (Gemsbok)	*Oryx gazella gazella*
Scimitar-horned oryx	*Oryx dammah*
OX	
Musk ox	*Ovibos moschatus*
PADEMELON	
Cape York pademelon	*Thylogale stigmatica coxeni*
PLATYPUS	*Ornithorhynchus anatinus*
POLECAT	
Striped polecat (Zorilla)	*Ictonyx striatus*
POSSUM	
Mountain pygmy possum	*Burramys parvus*
PRAIRIE DOG	*Cynomys ludovicianus*
PRZEWALSKI'S HORSE	*Equus przewalskii*
PUDU	*Pudu pudu*
PUMA	*Felis concolor*
QUAGGA	*Equus quagga*
RABBIT	*Oryctolagus cuniculus*
RACCOON	*Procyon lotor*
RAT	
Crested rat	*Lophiomys imhausi*
RATEL (Honey badger)	*Mellivora capensis*
REINDEER	*Rangifer tarandus*
RHINOCEROS	
Black rhinoceros	*Diceros bicornis*
Indian rhinoceros	*Rhinoceros unicornis*
White (Square-lipped) rhinoceros	*Ceratotherium simum*
Sumatran (Asian) rhinoceros	*Rhinoceros sumatrensis*
SAIGA	*Saiga tatarica*
SEAL	
Cape fur seal	*Arctocephalus pusillus*
Crab-eating seal	*Lobodon carcinophaga*
Southern elephant seal	*Mirounga leonina*
SEROW	
Japanese serow	*Capricornis crispus*

SHEEP	
Barbary sheep (Aoudad)	*Ammotragus (Capra) lervia*
Dall's sheep	*Ovis dalli*
Desert bighorn sheep	*Ovis canadensis mexicana*
SITATUNGA	*Tragelaphus spekeri*
SKUNK	*Mephitis mephitis*
SLOTH	
Two-toed sloth	*Choloepus didactylus*
SPRINGBOK	*Antidorcas marsupialis*
STAG	
Manchurian stag	*Cervus elaphus xanthopygus*
STEINBOK	*Raphicerus campestris*
TAHR	
Himalayan tahr	*Hemitragus jemlahicus*
TAKIN	*Budorcas taxicolor*
TAPIR	
Brazilian tapir	*Tapirus terrestris*
Baird's tapir	*Tapirus bairdi*
Malayan tapir	*Tapirus indicus*
TARPAN (Wild horse)	*Equus ferus*
TENREC	
Pygmy tenrec	*Echinops telfairi*
Spiny (Hedgehog) tenrec	*Setifer setosus*
THYLACINE (Tasmanian wolf)	*Thylacinus cynocephalus*
TIGER	
Bengal tiger	*Panthera tigris tigris*
Siberian (Amur) tiger	*Panthera tigris altaica*
South China tiger	*Panthera tigris amoyensis*
Sumatran tiger	*Panthera tigris sumatrae*
TUR	
West Caucasian tur	*Capra ibex caucasica*
East Caucasian tur	*Capra cylindricornis*
VONTSIRAFOTSY	*Galidictis striata*
WALLABY	
Bennett's wallaby	*Macropus rufogriseus*
Sandy wallaby	*Macropus agilis*
White-throated (Parma) wallaby	*Macropus parma*

WALRUS	*Odobenus rosmarus*
WAPITI (Red deer)	*Cervus elaphus*
WARTHOG	*Phacochoerus aethiopicus*
WILDCAT	
European wildcat	*Felis silvestris*
WILDEBEEST	
Black wildebeest	*Connochaetes gnou*
WISENT (European bison)	*Bison bonasus*
WOLF	
Maned wolf	*Chrysocyon brachyurus*
Timber wolf	*Canis lupus*
YAK	*Bos mutus*
ZEBRA	
Grant's zebra	*Equus burchelli boehmi*
Grevy's zebra	*Equus grevyi*
ZEBU	*Bos indicus*

Index

INDEX

INDEX

INDEX

INDEX

INDEX

For Product Safety Concerns and Information please contact our EU representative GPSR@taylorandfrancis.com Taylor & Francis Verlag GmbH, Kaufingerstraße 24, 80331 München, Germany

Batch number: 10725939

Printed by Printforce, the Netherlands